Jim Donald

Math
Principles
for Food
Service
Occupations

This book is dedicated to my daughter,
Mrs. Connie Kruetzkamp,
for her diligent work in typing the original manuscript.

MATH PRINCIPLES FOR FOOD SERVICE OCCUPATIONS

Second Edition

Robert G. Haines

 Delmar Publishers Inc.®

NOTICE TO THE READER

Publisher does not warrant or guarantee any of the products described herein or perform any independent analysis in connection with any of the product information contained herein. Publisher does not assume, and expressly disclaims, any obligation to obtain and include information other than that provided to it by the manufacturer.

The reader is expressly warned to consider and adopt all safety precautions that might be indicated by the activities described herein and to avoid all potential hazards. By following the instructions contained herein, the reader willingly assumes all risks in connection with such instructions.

The publisher makes no representations or warranties of any kind, including but not limited to, the warranties of fitness for particular purpose or merchantability, nor are any such representations implied with respect to the material set forth herein, and the publisher takes no responsibility with respect to such material. The publisher shall not be liable for any special, consequential or exemplary damages resulting, in whole or in part, from the readers' use of, or reliance upon, this material.

Delmar Staff

Managing Editor: Barbara A. Christie
Production Editor: Ruth East
Production Coordinator: Linda Helfrich
Art Production Coordinator: Linda Johnson

For information, address Delmar Publishers Inc.
2 Computer Drive West, Box 15-015
Albany, New York 12212

COPYRIGHT © 1988
BY DELMAR PUBLISHERS INC.

All rights reserved. Certain portions of this work copyright © 1979. No part of this work covered by the copyright hereon may be reproduced or used in any form or by any means—graphic, electronic, or mechanical, including photocopying, recording, or information storage and retrieval systems—without written permission of the publisher.

10 9 8 7 6 5 4 3

Printed in the United States of America
Published simultaneously in Canada
by Nelson Canada,
a division of International Thomson Limited

Library of Congress Cataloging in Publication Data

Haines, Robert G.
   Math principles for food service occupations/Robert G. Haines—
2nd ed.
     p. cm.
     Includes index.
     ISBN 0-8273-3131-2 (textbook).   ISBN 0-8273-3132-0 (instructor's
guide)
     1. Food service—Mathematics.   I. Title.
TX911.3.M33H35   1988   87-36491
513′.93′024642—dc19   CIP

Contents

Contents

PART FIVE: FOR ONE AND ALL

Preface

Mathematics is such an important part of any food service operation that it is necessary for all employees to have an understanding of the basic math concepts involved. The understanding of the math procedures that help show a business profit are often the key to advancement. It takes math skills to control food and labor costs—the two expenditures that determine a profit or loss.

Since its first printing, *Math Principles for Food Service Occupations*, has steadily increased in acceptance, making a second edition possible. All the information and concepts of the first edition were retained. In addition, mathematical topics have been added to create a more complete food service math workbook and a more effective teaching tool.

Math Principles for Food Service Occupations opens with a pre-test and concludes with a post-test for the purpose of evaluating the student's math skills before and after the course in order to determine the extent of student improvement. The pre- and post-test consist of twenty-five different math skills. A profile sheet for both the pre- and post-test can be kept on file for reference.

The pre-test is followed by the content, which is divided into five coordinated parts to simplify learning. Part one is a review intended to refresh and sharpen the student's math skills. The emphasis is placed on the methods used to solve mathematical problems related to food service situations. This information should be thoroughly reviewed, and the exercise problems worked and referred back to whenever necessary during the study of subsequent parts of the text.

Parts two, three and four deal with mathematical topics that apply directly to the operation of a food service establishment. Part two, which has been expanded from six units to nine units, focuses on the math necessary to function as part of the preparation crew. This part includes weights and measures, portion control, standard recipe conversion, daily food production reports, back of the house business forms, production formulas, balancing formulas, and finding approximate recipe yields and meat cut percentages. Part three involves the math for the service crew, dealing with sales checks, tipping, and the cashier's daily worksheet. Information has been added on how to use electronic calculators. Part four concerns the math for management procedures—figuring standard recipe costs, pricing the menu, writing daily food cost reports, doing inventories and financial statements, and a completely new unit concerning budgeting. Part five presents related math information that affects both the food service business and individual everyday existence. The topics discussed here are taxes, the metric system, checking accounts, and the important addition of simple interest and compound interest.

It is the Author's hope that the material contained in this book will provide the student with enough mathematical knowledge to allow rapid advancement to a position of management after the apprenticeship years have been completed.

ABOUT THE AUTHOR

Robert G. Haines is currently the chef of the executive dining room at the Cincinnati Milacron Company. He retired in 1985 after teaching 32 years at Norwood High School and Scarlet Oaks Career Development Center. For more than 40 years, Mr. Haines has been actively involved in different facets of the food service industry, having worked as both a teacher and a chef. During his career, the author has participated in several professional organizations and has served on many committees for the promotion and development of the food service industry. Among his accomplishments, Mr. Haines started the first food service program in a high school in the state of Ohio. He was called to Washington, D.C., by the Department of Labor to explore ways of increasing food service training programs throughout the United States. His experience in the trade has provided him with extensive knowledge of the subject and of students, as reflected in the practical, thorough approach to food service math presented in this text.

Acknowledgments

The author would like to thank the following for their contributions of content and illustration.

Gregory Ptacin, Promotion Director, The Cincinnati Post
Robert L. Bennett, Manager, Executive Dining Room, Citibank, NA
David Darby, Chef Instructor, Scarlet Oaks Career Development Center
Scovill Manufacturing Co., Hamilton Beach Division
United States Department of Commerce, National Bureau of Standards
American Hotel and Motel Association
National Cash Register Company
Proctor and Gamble Company
Raymond Sisson, Communications Coordinator, Great Oaks Joint Vocational
    School
Penn Scale Company
Lillie Bell Tyler, Home Economics Supervisor, Great Oaks Joint Vocational
    School

Other help was provided by:

G. C. Wilson, Ohio District Manager, Burger Chef Systems, Inc.
Rosemary Marshall, Counselor, Scarlet Oaks Career Development Center
Edward Hasenour, Hasenour's Restaurant
William Fisher, Executive Vice President, National Restaurant Association
John Humphrey, President, Zino's Restaurant
T. D. Hughes, General Manager, LaRosa's Restaurant

The author would also like to express his appreciation to his wife, Dolores, and his daughter, Connie, for their help in typing the original manuscript.

The instructional material in this text was classroom tested at the Scarlet Oaks Career Development Center, Sharonville, Ohio.

Pre-Test: Basic Math Skills

The pre-test evaluates a student's math skills before the student starts the food service math course. It helps the student and teacher focus on areas of greatest concern.

To earn a competency in each of the twenty-five math skills presented, a student must work three of the four problems presented correctly. If this is achieved, the student will earn a + (plus) for that particular math skill. If this goal is not achieved, a − (minus) will be recorded. A profile sheet on both the pre- and post-tests are kept on file by the teacher for reference by either the student or the teacher. The pluses are recorded in either blue or black ink on the profile sheet under the proper math skill. The minuses are recorded in red ink.

1. Add the following numbers.

$$27 \quad\quad 17 \quad\quad 6555$$
$$+49 \quad\quad 46 \quad\quad 2265$$
$$\quad\quad\quad 13 \quad\quad 8085$$

$$38 + 127 + 5,678 + 42,975 =$$

2. Subtract the following numbers.

$$48 \quad\quad 658 \quad\quad 82,723$$
$$-26 \quad\quad -96 \quad\quad -3,430$$

$$34,228 - 16,767 =$$

3. Change the mill to the nearest cent.

$$\$.034 \quad\quad\quad \$6.857$$
$$\$.126 \quad\quad\quad \$725.683$$

4. Multiply the following numbers.

$$54 \quad\quad 68 \quad\quad 981$$
$$\times 9 \quad\quad \times 45 \quad\quad \times 127$$

$$2520 \times 36 =$$

5. Divide the following numbers.

$$12\overline{)192} \quad\quad 18\overline{)1350}$$

$$122\overline{)76,860} \quad\quad 2125\overline{)518,750}$$

6. Reduce the fractions to the lowest terms.

$$\frac{4}{12} \quad \frac{16}{96} \quad \frac{56}{64} \quad \frac{60}{108}$$

7. Convert each mixed number to an improper fraction.

$$6\frac{1}{8} \quad\quad 16\frac{3}{7}$$

$$7\frac{3}{8} \quad\quad 41\frac{5}{8}$$

8. Convert each improper fraction to a whole number or to a mixed number.

$$\frac{41}{9} \quad\quad\quad \frac{192}{64}$$

$$\frac{28}{6} \quad\quad\quad \frac{270}{24}$$

9. Find the equivalent fractions.

$$\frac{5}{6} = \frac{?}{30} \quad\quad \frac{6}{15} = \frac{?}{60}$$

$$\frac{5}{9} = \frac{?}{36} \quad\quad \frac{2}{3} = \frac{?}{21}$$

10. Add the following fractions and reduce the answer to the lowest terms.

$$\frac{3}{7} \quad\quad 3\frac{7}{8} \quad\quad 1\frac{3}{16} \quad\quad 2\frac{1}{2}$$

$$+\frac{2}{7} \quad\quad +1\frac{1}{4} \quad\quad +4\frac{3}{4} \quad\quad 4\frac{3}{4}$$

$$\quad\quad\quad\quad\quad\quad\quad\quad\quad\quad\quad\quad +8\frac{1}{12}$$

11. Subtract the following fractions and reduce the answer to the lowest terms.

$$\frac{9}{12} \qquad 8 \qquad 12\frac{1}{4} \qquad 42\frac{23}{32}$$
$$-\frac{5}{12} \qquad -3\frac{5}{8} \qquad -6\frac{5}{16} \qquad -26\frac{15}{16}$$

12. Multiply the following fractions and reduce the answer to the lowest terms.

$$\frac{3}{5} \times \frac{5}{6} = \qquad\qquad 1\frac{3}{4} \times 2\frac{3}{8} =$$

$$8 \times \frac{4}{9} = \qquad\qquad 8\frac{2}{3} \times 2\frac{1}{4} =$$

13. Divide the following fractions and reduce the answer to the lowest terms.

$$\frac{2}{9} \div \frac{1}{3} = \qquad\qquad 1\frac{3}{4} \div \frac{2}{3} =$$

$$1\frac{1}{4} \div \frac{3}{8} = \qquad\qquad 12\frac{3}{4} \div \frac{1}{3} =$$

14. Convert each decimal to a fraction.

$$.7 \qquad\qquad .009$$
$$4.15 \qquad\qquad .89$$

15. Convert each fraction to a decimal.

$$\frac{3}{10} \qquad\qquad 5\frac{1}{6}$$

$$\frac{3}{8} \qquad\qquad 14\frac{5}{8}$$

16. Convert each fraction to a percent.

$$\frac{3}{5} \qquad\qquad \frac{7}{8}$$

$$1\frac{2}{5} \qquad\qquad \frac{3}{10}$$

17. Convert each decimal to a percent.

$$23.4 \qquad\qquad .67$$
$$.0065 \qquad\qquad 49.3$$

18. Convert each percent to a decimal.

$$8.5\% \qquad\qquad 12.6\%$$
$$228\% \qquad\qquad 25\%$$

19. Add the decimals.

$$9.6 \qquad\qquad 8.085$$
$$4.6 \qquad\qquad +\ 12$$
$$+\ .8$$

$$.6 + 8.4 + 10 =$$
$$.9946 + .023 + .0425 =$$

20. Subtract the decimals.

$$9.08 \qquad\qquad 221.06$$
$$-\ 3.57 \qquad\qquad -9.2$$

$$42.3 - 10.63 =$$
$$8 - .04 =$$

21. Multiply the decimals.

$$.275 \qquad\qquad 7.38$$
$$\times\ \ 15 \qquad\qquad \times\ 2.9$$

$$6.5 \times .043 =$$
$$10.85 \times .034 =$$

22. Divide the decimals.

$$.06\overline{).855} \qquad\qquad 9.5\overline{).4832}$$

$$.44\overline{)5853} \qquad\qquad 18\overline{)9.683}$$

23. Use percents to find the total number.

20% of what number is 10?
18% of what number is 36?
60% of what number is 45?
40% of what number is 20?

24. Find the following numbers.

    36% of 8 =

    6.8% of 36 =

    112% of 200 =

    $14\frac{1}{2}$% of 85 =

25. Find the following percentages.

    What percent of 40 is 20?

    What percent of 200 is 88?

    What percent of 150 is 90?

    What percent of 360 is 126?

Basic Math Skills—Profile Sheet

+ Mastered
− Not Mastered

Student's Name	Addition	Subtraction	Mills	Multiplication	Division	Reducing fractions	Converting to improper fraction	Improper fractions to mixed number	Finding equivalent fractions	Addition of fractions	Subtraction of fractions	Multiplication of fractions	Division of fractions	Converting decimals to fractions	Converting fractions to decimals	Converting fractions to percent	Converting decimals to percent	Converting percent to decimals	Addition of decimals	Subtraction of decimals	Multiplication of decimals	Division of decimals	Percents to find number	Finding percent of a number	Finding percent of two given numbers
1.																									
2.																									
3.																									
4.																									
5.																									
6.																									
7.																									
8.																									
9.																									
10.																									
11.																									
12.																									
13.																									
14.																									
15.																									
16.																									
17.																									
18.																									
19.																									
20.																									
21.																									
22.																									
23.																									
24.																									
25.																									

PART ONE

Review
of Basic Math

This section contains a review of the basic math that is used in food service operation. Whole numbers, fractions, decimals, and percents are some of the topics covered. Most of the math connected with food service operations is quite simple. An understanding of this math is essential for giving accuracy to record keeping, increasing the success of the business, and adding interest to a food service career.

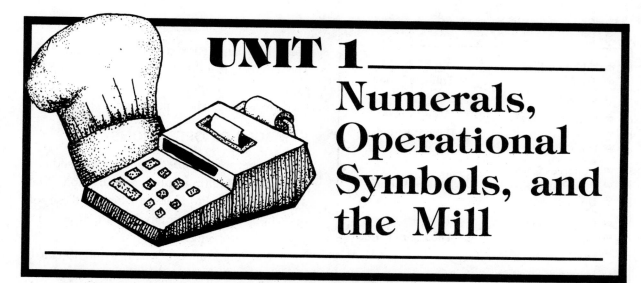

UNIT 1

Numerals, Operational Symbols, and the Mill

The material contained in this section is simple but important to the study of food service operations. First, it is necessary to understand the meaning of certain math terms, such as whole number, unit, numeral, and digit.

Whole numbers are numbers such as 0, 1, 2, 3, etc. which are used to represent whole units rather than fractional units. A *unit* is a single quantity of like things, such as one case of soda. A unit can also be a single item, such as a side of beef or a pork loin. Units are sometimes established by producers or manufacturers. The unit established is usually a quantity which is convenient for both the consumer and the producer. For example, packaging soda in 1-ounce cans would not satisfy most people. Likewise, if the soda was only available in 200-gallon vats, the quantity would be difficult to drink, and would therefore be impractical. Sensible units also help a manufacturer keep track of inventory. Units, of course, are not limited to manufactured products. A unit is, however, always a single quantity of like things.

Numerals are used to represent numbers.

Digits are any of the numerals that combine to form numbers. There are ten digits, as shown in figure 1-1, with their names.

Digits can be combined in different ways to produce different numbers. For example, 436 and 643 are both combinations of the digits 4, 3, and 6. The value of each digit depends on its location (or *place*) in the number. Each place has a different name and a different value, as shown in figure 1-2.

In large numbers, made up of four or more digits, the digits are put into groups of three. Each of these groups is called a *period,* as shown in figure 1-2. Periods are separated with commas. (In the metric system, to be described later, periods are separated by space, rather than commas.) The value of each digit is determined by its position in the place value columns. In figure 1-2, the digit 5 is used twice. When it appears

0	1	2	3	4	5	6	7	8	9
zero	one	two	three	four	five	six	seven	eight	nine

Fig. 1-1 Digits

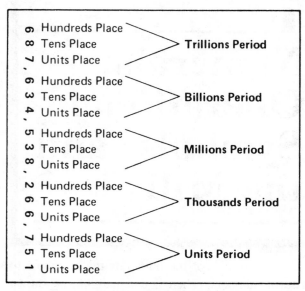

Fig. 1-2 Place value columns within each period.

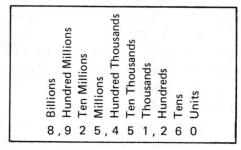

Fig. 1-3 Place values.

in the tens place of the units period, the 5 represents 50. When it is used in the hundreds place of the millions period, it represents 500 million. Place, therefore, is very important.

Large numbers are also expressed with the names shown in figure 1-3. The digit zero (0) in the units column is needed to hold a place and to give the other digits their proper value. The digit 6 in the tens place would be 6, not 60, without the zero. The complete number shown in figure 1-3 is read eight billion, nine hundred twenty-five million, four hundred fifty-one thou-

sand, two hundred sixty. Zeros are not read. The number 129,000,000 is read one hundred twenty-nine million. The word "and" is not used in reading whole numbers.

SYMBOLS OF OPERATIONS

There are four basic operations: addition, subtraction, multiplication, and division. Math symbols are used to indicate which of these four operations is needed in any given transaction or arithmetic problem.

The importance of math symbols can be quickly demonstrated. When the same two numerals are used with each of the four math symbols, they can yield quite different results:

$$
\begin{array}{llll}
\text{a.} \quad \begin{array}{r} 8 \\ + 4 \\ \hline 12 \end{array} & \text{b.} \quad \begin{array}{r} 8 \\ \times 4 \\ \hline 32 \end{array} & \text{c.} \quad \begin{array}{r} 8 \\ - 4 \\ \hline 4 \end{array} & \text{d.} \quad 8 \div 4 = 2
\end{array}
$$

Example a, of course, is addition, b is multiplication, c is subtraction, and d is division. Figure 1-4 gives the names, meanings, and some examples of the symbols commonly used in food service.

THE MILL

When dealing with monetary numbers, the word *cent* is used to represent the value of one hundredth part of a dollar. The third place to the right of the decimal is called a *mill* and represents the thousandth part of a dollar, or one tenth of one cent.

When the final result of a monetary number includes a mill, it is usually rounded to a whole number of cents. To round a number to the nearest cent, the third digit (mill) is dropped, but if that digit is 5 or more, add another cent to the digit before it.

SYMBOL	NAME	MEANING	EXAMPLES
+	plus sign	add to, or increased by	8 + 99 = 107 2 + 19 = 21 361 + 12 = 373
−	minus sign	subtract from, take away from, decreased by, or less	17 − 9 = 8 23 − 5 = 18 49 − 9 = 40
×	multiplication or times sign	multiply by, or the product of	2 × 12 = 24 9 × 3 = 27 5 × 25 = 125
÷	division sign	divided by	4 ÷ 2 = 2 27 ÷ 9 = 3 100 ÷ 20 = 5
—	fraction bar	also represents division	$\frac{6}{3} = 2$ $\frac{10}{5} = 2$ $\frac{20}{4} = 5$
·	decimal point	indicates the beginning of a decimal fraction	0.321 1.877 117.65
%	percent sign	parts per 100, by the hundredths	15% 12% 6%
=	equal sign	the same value as, or is equal to	1 = 1
$	dollar sign	the symbol placed before a number to indicate that it stands for dollars	$12.00
@	at or per	used to indicate price or weight of each unit when there is a quantity of a unit	5 dz. doughnuts @$1.15/dz. 25 bags of potatoes @ 10 lb./bag 100 ice cream cones @ $.25 per cone

Fig. 1-4 **Mathematical symbols.**

Examples:

$472.237 rounded to the nearest cent is $472.24, because 7 mills are more than 5.

$472.234 rounded to the nearest cent is $472.23, because 4 mills are less than 5.

In different businesses, the method of rounding may vary. In some cases, the amount is rounded to the nearest cent; in others, any fraction of a cent is changed to a whole cent. In still other cases, the fourth digit is also considered. When solving the problems in this text, figure only to the third digit to the right of the decimal, which is the mill.

ACHIEVEMENT REVIEW
THE MILL

A. Answer the following questions about the mill.
1. How many mills are contained in 1 cent?
2. How many mills are contained in 10 cents?
3. How many mills are contained in ½ cent?
4. How many mills are contained in $1.00?
5. What is the rule to follow if the mill is 4 or less? Five or more?

B. Change to the nearest cent.

6. $.025
7. $.034
8. $.46
9. $.057
10. $.038
11. $.052
12. $.063
13. $.083
14. $.094
15. $.066

16. $.013
17. $.039
18. $.106
19. $.114
20. $.125
21. $648.524
22. $8,920.578
23. $9,450.683
24. $724.255
25. $549.239

UNIT 2
Addition, Subtraction, Multiplication, and Division

ADDITION

Addition is simply combining things or units which are alike to find the sum or total. If a person has $4.00 and is given $10.00, the person has a total of $14.00. This simple addition problem can be written $4.00 + $10.00 = $14.00. Placing the numbers in a row is called the horizontal position. The horizontal position is seldom used in a problem involving large numbers because it is difficult to calculate the answer. The money addition can also be written:

$$
\begin{array}{r}
\$\ 4.00 \\
+\ 10.00 \\
\hline
\$14.00
\end{array}
$$

This is called the vertical or column position. The plus sign is used to indicate that the numbers are to be added. Figure 2-1, page 9, shows an example of how addition is used in food service.

For another example: If a fast food service operation has 117 chickens stored in the cooler, 12 chickens deep frying, and 21 chickens ready to serve, the food establishment has a total of 150 chickens on hand. Since all of the units to be added are alike, it is unnecessary to write out what the units are. Therefore, the addition can be written in either of the following ways:

$$117 + 12 + 21 = 150$$

or

$$
\begin{array}{r}
117 \\
12 \\
21 \\
\hline
150
\end{array}
$$

The plus sign is not used when three or more numbers are added together in the vertical position because it cannot be confused with any other arithmetic operation. Remember that each digit must be placed correctly to give it the proper value.

There are several basic tips that can be followed in arithmetic problems to save time and to improve accuracy.

Be neat. For the person interested in a food service career, neatness is essential in both physical appearance and personal hygiene. Neatness is also important in math. It does not require much time to write neatly and carefully, placing each number in the proper column directly under the number above it.

```
 2034
   21
  516
  207
   65
 1903
 4746
```

Neatness is also important when carrying numbers over from one column to the next. The sum of the units column in the preceding problem is 26 and not 6. The 2 (actually 20) is carried over to the tens column. If the carryover 2 is written, it is placed neatly at the top of the tens column, so it is not overlooked when counting that column. The sum of the tens column is 14 (actually 140). The 1 is carried over to the hundreds column. Again, place the 1 neatly over the top of the hundreds column. Follow this procedure for each column in the problem.

```
       ①①②
      2034
        21
       516
       207
        65
      1903
      4746
```

Check all work. All work must be checked to be sure the addition is correct. If it is found, through checking, that the work is always correct, this practice should still be continued. The penalty for mathematical mistakes in the classroom is a lower grade; the penalty for these mistakes in a food service operation can result in a monetary loss for both the employee and the employer.

The common method of checking addition is to add the individual columns in reverse order. For example, if the units column is first added

from bottom to top, check the work by adding again, this time from top to bottom.

Increase accuracy and speed. Addition is often simplified if numbers are combined and then added. For example, in the problem $8 + 2 + 6 + 4 + 4 + 2 = 26$, the addition is greatly simplified by combining 8 and 2 into 10, and 6 and 4 into 10, which adds up to twenty. Adding the remaining numbers (4 and 2) gives 6, which makes the total 26.

Eliminate unnecessary words when adding. Instead of saying eight plus two equals ten, automatically see the eight and two combination as ten. When adding the problem in the previous paragraph, don't say ten plus ten equals twenty, plus four equals twenty-four, plus two equals twenty-six. Say ten, twenty, twenty-six.

Another method of increasing speed and accuracy is to total each column as if it is a separate problem. The sum of each column is placed on a separate line and then these sums are totaled. This method eliminates the need to carry over numbers to the next column.

thousands	hundreds	tens	units	
2	6	5	4	
3	1	8	7	
6	1	2	1	
8	2	2	9	
4	8	7	1	
9	7	0	3	
		2	5	sum of units column
	2	4	0	sum of tens column
2	5	0	0	sum of hundreds column
3 2	0	0	0	sum of thousands column
3 4	7	6	5	sum total of problem

Fig. 2-1 How many servings of Baba au Rhum cakes? This is an excellent example of the use of addition in food service.

It is not necessary to write the zeros which appear in the subtotals. They are shown here to illustrate that the sum of the tens column, for example, is 240 and not 24; the sum of the hundreds column is 2500 and not just 25; and so on. Writing out the zeros makes it possible to place all digits in their proper columns. This method of addition usually starts in the units column, as shown above, however, the same results are obtained by starting the adding from the left column. In the above problem, this means beginning the addition in the thousands column.

The following is an example of this approach.

thousands	hundreds	tens	units	
2	6	5	4	
3	1	8	7	
6	1	2	1	
8	2	2	9	
4	8	7	1	
9	7	0	3	
3 2	0	0	0	sum of thousands column
2	5	0	0	sum of hundreds column
	2	4	0	sum of tens column
		2	5	sum of units column
3	4	7	6 5	sum of total of problem

Whichever approach is chosen, all work should be checked.

If these tips are followed, it will help to ensure the accuracy and speed required for any type of addition problem, especially those related to food service, figure 2-1. Some of these tips, such as the need for neatness, apply to all arithmetic operations.

ACHIEVEMENT REVIEW
ADDITION

Find the sum of each of the following addition problems. Use the methods and tips suggested in the text. Check all work.

1. 7
  +8

2. 5
  +9

3. 2
    6
    9
    4

4. 8
    9
    7
    6

5. 9
    4
    5
    3

6. 2 + 4 + 7 + 9 =
7. 9 + 8 + 7 + 12 =
8. 20 + 18 + 16 + 14 =
9. 32 + 28 + 26 + 15 =
10. 64 + 24 + 46 + 52 =

11. 249
   +148

12. 546
   +417

13. 415
     325
     640

14. 649
     538
     427

15. $49.18
    43.56
    28.37
    18.55

16. 410
    522
    344
    238
    656

17. $22.48
    20.56
    18.49

18. $39.18
    44.56
    28.47
    18.54

19. $468.29
    549.32
    376.51
    225.25
    152.46

20. $979.29
    718.26
    631.33
    844.23
    451.19
    323.76

21. 8
    57
    348
    2876
    420
    566
    18
    19
    249
    302

22. 4418
    4318
    4217
    3921
    3822
    3772
    2552
    2642
    2225
    4345

23. $28.43
    54.56
    42.25
    35.67
    27.23
    17.19
    10.42
    52.39
    60.25
    71.41

24. $542.23
    218.12
    140.19
    410.20
    612.13
    716.15
    240.41
    321.22
    125.25
    444.44

25. $ 1.43
    2.45
    5.22
    4.16
    3.17
    7.28
    6.29
    8.36
    9.42
    10.13

SUBTRACTION

Subtraction is the removal of one number of things from another number of things, figure 2-2. It is often thought of as the opposite of addition.

If a person has $10.00 and spends $9.38, the subtraction problem is written as follows:

$$\begin{array}{r} \$10.00 \\ -\ 9.38 \\ \hline \$\ .62 \end{array}$$

The minus sign ($-$) must always be used so that the problem is not confused with another operation.

Each of the factors in subtraction has a name. The original number before the subtraction takes place is called the *minuend*. In the above problem, the minuend is $10.00. The number removed from the minuend is called the *subtrahend*. Finally, the amount left over or remaining after the problem is completed is called the *remainder* (or *difference*).

$$\begin{array}{rl} 1492 & \text{minuend} \\ -\ 701 & \text{subtrahend} \\ \hline 791 & \text{remainder or difference} \end{array}$$

Borrowing

Subtraction frequently requires borrowing. Although the minuend is usually a larger number than the subtrahend, a particular digit in the

Fig. 2-2 Subtraction is taking place when a serving portion is removed from a whole unit.

minuend may be less than the digit beneath it in the subtrahend. Example: $1624 - 798 = 826$. If this problem is set up in the vertical position, it looks like this:

```
thousands
  hundreds
    tens
      units

1   6   2   4    minuend
  – 7   9   8    subtrahend
    8   2   6    remainder or difference
```

The minuend (1624) is clearly a larger number than the subtrahend (798). However, the digit 8 in the units column of the subtrahend is larger than the digit 4 in the units column of the minuend. Since 8 cannot be subtracted from 4, it is necessary to borrow from the tens column.

To indicate the fact that a ten has been borrowed, cross out the 2 in the tens column and write 1 above it.

```
          1
1   6   2̸ ¹4
  – 7   9   8
```

Add the borrowed ten to the four in the units column, which gives 14. Then subtract: $14 - 8 = 6$. The six is written beneath the bar in the units column.

```
          1
1   6   2̸ ¹4
  – 7   9   8
              6
```

In the tens column, 9 cannot be subtracted from 1 (actually 90 from 10), so it is necessary to borrow a hundred from the hundreds column. This is done by crossing out the numeral 6 in the hundreds column and writing 5 above it.

```
      5   11
1   6̸   2̸ ¹4
  – 7   9   8
              6
```

Returning to the tens column, subtract 9 from 11 (actually 90 from 110), to get 2 (actually 20). Write the 2 in the tens column beneath the bar.

```
      5   11
1   6̸   2̸ ¹4
  – 7   9   8
          2   6
```

Moving to the hundreds column, 7 cannot be subtracted from 5 (which is actually 700 from 500), so one thousand is borrowed from the thousands column. Cross out the numeral 1 in the thousands column.

```
      15  11
1   6̸   2̸   4
  – 7   9   8
          2   6
```

Subtract 7 from 15 (actually 700 from 1500), to get 8 (actually 800). Write the eight in the hundreds column beneath the bar. The completed problem looks like this:

```
      15  11
1̸6̸   2̸ ¹4
  – 7   9   8
      8   2   6
```

Checking

The common way of checking an answer in subtraction is to add together the subtrahend and the remainder. The sum of these two numbers should equal the minuend.

Subtraction		Check	
2 6 7 4	minuend	1 9 3 7	subtrahend
− 1 9 3 7	subtrahend	+ 7 3 7	remainder
7 3 7	remainder	2 6 7 4	minuend

ACHIEVEMENT REVIEW
SUBTRACTION

Find the difference in the following subtraction problems. Check your work.

1. 945 − 203 =

2. 876 − 545 =

3. 749 − 438 =

4. 2049 − 523 =

5. 3955 − 1542 =

6.
$$
\begin{array}{r}
842 \\
-421 \\
\hline
\end{array}
$$

7.
$$
\begin{array}{r}
928 \\
-617 \\
\hline
\end{array}
$$

8.
$$
\begin{array}{r}
1018 \\
-1005 \\
\hline
\end{array}
$$

9.
$$
\begin{array}{r}
4197 \\
-2076 \\
\hline
\end{array}
$$

10.
$$
\begin{array}{r}
746 \\
-458 \\
\hline
\end{array}
$$

11.
$$
\begin{array}{r}
958 \\
-269 \\
\hline
\end{array}
$$

12.
$$
\begin{array}{r}
2049 \\
-1659 \\
\hline
\end{array}
$$

13.
$$
\begin{array}{r}
\$27.53 \\
-12.78 \\
\hline
\end{array}
$$

14.
$$
\begin{array}{r}
\$59.45 \\
-32.79 \\
\hline
\end{array}
$$

15.
$$
\begin{array}{r}
\$67.42 \\
-58.53 \\
\hline
\end{array}
$$

16.
$$
\begin{array}{r}
\$128.76 \\
-57.98 \\
\hline
\end{array}
$$

17.
$$
\begin{array}{r}
\$247.98 \\
-159.67 \\
\hline
\end{array}
$$

18.
$$
\begin{array}{r}
\$398.20 \\
-177.59 \\
\hline
\end{array}
$$

19.
$$
\begin{array}{r}
\$5,429.28 \\
-1,336.79 \\
\hline
\end{array}
$$

20.
$$
\begin{array}{r}
\$6,222.43 \\
3,134.64 \\
\hline
\end{array}
$$

21.
$$
\begin{array}{r}
8,962,394 \\
-7,350,272 \\
\hline
\end{array}
$$

22.
$$
\begin{array}{r}
\$57,620.00 \\
-40,232.75 \\
\hline
\end{array}
$$

23.
$$
\begin{array}{r}
\$74,769.00 \\
-38,410.12 \\
\hline
\end{array}
$$

24.
$$
\begin{array}{r}
91,313,802 \\
-35,286,805 \\
\hline
\end{array}
$$

25.
$$
\begin{array}{r}
\$121,002.02 \\
-73,624.05 \\
\hline
\end{array}
$$

MULTIPLICATION

Multiplication can be thought of as a shortcut for a certain type of addition problem. In multiplication, a whole number is added to itself a specified amount of times. $3 \times 2 = 6$ is another way of expressing $3 + 3 = 6$. The number 3 is added to itself two times.

Because of the relationship between addition and multiplication, it does not matter which operation is used in very simple problems, such as the one above. However, when faced with a problem involving large numbers (such as $4633 \times 9718 = ?$), addition is a very impractical way of solving the problem. This is when multiplication is useful as a shortcut for addition.

Each number involved in multiplication has a name. The number which is added to itself (3 in the first example) is called the *multiplicand* (which means "going to be multiplied"). The number representing the amount of times the multiplicand is to be added to itself is called the *multiplier* (number 2 in the first example). The result of multiplying the multiplicand by the multiplier is called the *product*. The product in the first example is 6.

The following example gives the names and functions of the various numbers involved in a multiplication problem.

$$
\begin{array}{rl}
4\ 6\ 8 & \text{multiplicand} \\
\times\ 2\ 4 & \text{multiplier} \\
\hline
1\ 8\ 7\ 2 & \text{subproduct} \\
9\ 3\ 6\ 0 & \text{subproduct} \\
\hline
1\ 1{,}2\ 3\ 2 & \text{product}
\end{array}
$$

Function	Name
number to be added to itself	multiplicand
number of times to be added to itself	multiplier
product of the units column	subproduct
product of the tens column	subproduct
final product	product

In this example, subproducts are shown. Subproducts occur whenever the multiplier consists of two or more digits. In this case, the multiplier is 24. The first subproduct is the result of the product $468 \times 4 = 1872$. The second subproduct (9360) is the result of multiplying 468 times 20. The zero in this subproduct is not necessary because $2 + 0 = 2$. It does not affect the outcome of the problem. However, it is shown here to illustrate that the product of multiplying 468 times 20 is 9360 and not 936. It also helps keep all of the digits in their proper columns, (units in the units column, tens in the tens column, and so forth) as mentioned earlier in relation to subtraction.

Once all of the subproducts are determined, they are added together to obtain the final total or product (in this example, 11,232). The multiplication sign (also called the times sign) is always used in a multiplication problem, to distinguish it from any other type of arithmetic operation.

The Multiplication Table

It has been shown that multiplication is a shortcut for certain types of addition problems. A shortcut method is only valuable if it can be used efficiently and accurately. The key to using multiplication efficiently is the multiplication table, figure 2-3. (This gives products up to 12×12.)

Practice the table until it is memorized.

1 × 1 = 1			2 × 1 = 2			3 × 1 = 3	
1 × 2 = 2			2 × 2 = 4			3 × 2 = 6	
1 × 3 = 3			2 × 3 = 6			3 × 3 = 9	
1 × 4 = 4			2 × 4 = 8			3 × 4 = 12	
1 × 5 = 5			2 × 5 = 10			3 × 5 = 15	
1 × 6 = 6			2 × 6 = 12			3 × 6 = 18	
1 × 7 = 7			2 × 7 = 14			3 × 7 = 21	
1 × 8 = 8			2 × 8 = 16			3 × 8 = 24	
1 × 9 = 9			2 × 9 = 18			3 × 9 = 27	
1 × 10 = 10			2 × 10 = 20			3 × 10 = 30	
1 × 11 = 11			2 × 11 = 22			3 × 11 = 33	
1 × 12 = 12			2 × 12 = 24			3 × 12 = 36	
4 × 1 = 4			5 × 1 = 5			6 × 1 = 6	
4 × 2 = 8			5 × 2 = 10			6 × 2 = 12	
4 × 3 = 12			5 × 3 = 15			6 × 3 = 18	
4 × 4 = 16			5 × 4 = 20			6 × 4 = 24	
4 × 5 = 20			5 × 5 = 25			6 × 5 = 30	
4 × 6 = 24			5 × 6 = 30			6 × 6 = 36	
4 × 7 = 28			5 × 7 = 35			6 × 7 = 42	
4 × 8 = 32			5 × 8 = 40			6 × 8 = 48	
4 × 9 = 36			5 × 9 = 45			6 × 9 = 54	
4 × 10 = 40			5 × 10 = 50			6 × 10 = 60	
4 × 11 = 44			5 × 11 = 55			6 × 11 = 66	
4 × 12 = 48			5 × 12 = 60			6 × 12 = 72	
7 × 1 = 7			8 × 1 = 8			9 × 1 = 9	
7 × 2 = 14			8 × 2 = 16			9 × 2 = 18	
7 × 3 = 21			8 × 3 = 24			9 × 3 = 27	
7 × 4 = 28			8 × 4 = 32			9 × 4 = 36	
7 × 5 = 35			8 × 5 = 40			9 × 5 = 45	
7 × 6 = 42			8 × 6 = 48			9 × 6 = 54	
7 × 7 = 49			8 × 7 = 56			9 × 7 = 63	
7 × 8 = 56			8 × 8 = 64			9 × 8 = 72	
7 × 9 = 63			8 × 9 = 72			9 × 9 = 81	
7 × 10 = 70			8 × 10 = 80			9 × 10 = 90	
7 × 11 = 77			8 × 11 = 88			9 × 11 = 99	
7 × 12 = 84			8 × 12 = 96			9 × 12 = 108	
10 × 1 = 10			11 × 1 = 11			12 × 1 = 12	
10 × 2 = 20			11 × 2 = 22			12 × 2 = 24	
10 × 3 = 30			11 × 3 = 33			12 × 3 = 36	
10 × 4 = 40			11 × 4 = 44			12 × 4 = 48	
10 × 5 = 50			11 × 5 = 55			12 × 5 = 60	
10 × 6 = 60			11 × 6 = 66			12 × 6 = 72	
10 × 7 = 70			11 × 7 = 77			12 × 7 = 84	
10 × 8 = 80			11 × 8 = 88			12 × 8 = 96	
10 × 9 = 90			11 × 9 = 99			12 × 9 = 108	
10 × 10 = 100			11 × 10 = 110			12 × 10 = 120	
10 × 11 = 110			11 × 11 = 121			12 × 11 = 132	
10 × 12 = 120			11 × 12 = 132			12 × 12 = 144	

Fig. 2-3 Multiplication table of numbers from 1 to 12.

Then, to make sure it has been learned, write each problem on one side of an index card and the product on the other side. It should be possible to look at a problem and know the answer within 5 seconds, without looking at the other side.

Another method of presenting the multiplication table is shown in figure 2-4. This table gives the products of numbers up to 25 × 25 = 625. It is relatively simple to use. For example, to find the product of eight times nine, locate the number 8 in the vertical (up and down) column

to the far left. Then move your finger to the right until the nine is located in the horizontal (left to right) column at the top of the table. The number 72 is in the place where the eight column and the nine column intersect. Therefore, 72 is the product of 8 × 9. Look one place below the 72 and find the number 81. This is the product of 9 × 9. Drop down another place to find that 10 × 9 = 90. Using a card or sheet of paper across the table horizontally is helpful in locating the products of the various numbers.

1	2	3	4	5	6	7	8	9	10	11	12	13	14	15	16	17	18	19	20	21	22	23	24	25
2	4	6	8	10	12	14	16	18	20	22	24	26	28	30	32	34	36	38	40	42	44	46	48	50
3	6	9	12	15	18	21	24	27	30	33	36	39	42	45	48	51	54	57	60	63	66	69	72	75
4	8	12	16	20	24	28	32	36	40	44	48	52	56	60	64	68	72	76	80	84	88	92	96	100
5	10	15	20	25	30	35	40	45	50	55	60	65	70	75	80	85	90	95	100	105	110	115	120	125
6	12	18	24	30	36	42	48	54	60	66	72	78	84	90	96	102	108	114	120	126	132	138	144	150
7	14	21	28	35	42	49	56	63	70	77	84	91	98	105	112	119	126	133	140	147	154	161	168	175
8	16	24	32	40	48	56	64	72	80	88	96	104	112	120	128	136	144	152	160	168	176	184	192	200
9	18	27	36	45	54	63	72	81	90	99	108	117	126	135	144	153	162	171	180	189	198	207	216	225
10	20	30	40	50	60	70	80	90	100	110	120	130	140	150	160	170	180	190	200	210	220	230	240	250
11	22	33	44	55	66	77	88	99	110	121	132	143	154	165	176	187	198	209	220	231	242	253	264	275
12	24	36	48	60	72	84	96	108	120	132	144	156	168	180	192	204	216	228	240	252	264	276	288	300
13	26	39	52	65	78	91	104	117	130	143	156	169	182	195	208	221	234	247	260	273	286	299	312	325
14	28	42	56	70	84	98	112	126	140	154	168	182	196	210	224	238	252	266	280	294	308	322	336	350
15	30	45	60	75	90	105	120	135	150	165	180	195	210	225	240	255	270	285	300	315	330	345	360	375
16	32	48	64	80	96	112	128	144	160	176	192	208	224	240	256	272	288	304	320	336	352	368	384	400
17	34	51	68	85	102	119	136	153	170	187	204	221	238	255	272	289	306	323	340	357	374	391	408	425
18	36	54	72	90	108	126	144	162	180	198	216	234	252	270	288	306	324	342	360	378	396	414	432	450
19	38	57	76	95	114	133	152	171	190	209	228	247	266	285	304	323	342	361	380	399	418	437	456	475
20	40	60	80	100	120	140	160	180	200	220	240	260	280	300	320	340	360	380	400	420	440	460	480	500
21	42	63	84	105	126	147	168	189	210	231	252	273	294	315	336	357	378	399	420	441	462	483	504	525
22	44	66	88	110	132	154	176	198	220	242	264	286	308	330	352	374	396	418	440	462	484	506	528	550
23	46	69	92	115	138	161	184	207	230	253	276	299	322	345	368	391	414	437	460	483	506	529	552	575
24	48	72	96	120	144	168	192	216	240	264	288	312	336	360	384	408	432	456	480	504	528	552	576	600
25	50	75	100	125	150	175	200	225	250	275	300	325	350	375	400	425	450	475	500	525	550	575	600	625
1	2	3	4	5	6	7	8	9	10	11	12	13	14	15	16	17	18	19	20	21	22	23	24	25

Fig. 2-4 Multiplication table of numbers from 1 to 25.

Step-by-Step Multiplication

Another example of a multiplication problem clearly illustrates the step-by-step procedures involved in finding the product. As mentioned earlier, multiplication is a variation of addition. In the following example, it is shown that the product of the problem is the result of adding together the subproducts of each step of the problem.

Example: 913
          × 75

	thousands	hundreds	tens	units
Subproducts of Units Column				
Step 1. 5 × 3 =			1	5
Step 2. 5 × 10 =			5	0
Step 3. 5 × 900 =	4	5	0	0
Subproduct of units column	4	5	6	5

	thousands	hundreds	tems	units
Subproducts of Tens Column				
Step 4. 70 × 3 =		2	1	0
Step 5. 70 × 10 =		7	0	0
Step 6. 70 × 900 =	6 3,	0	0	0
Subproduct of tens column	6 3,	9	1	0

Step 7.   4565 + 63,910 = 68,475 Product

The above example showed the steps in finding the product of 913 × 75. Generally when performing a problem, the unneeded zeros are not written, but carryover numbers are used.

②   Carryover number for tens col.
①   Carryover number for units col.
913   Multiplicand
×   75   Multiplier
4565   Subproduct of units column
6391   Subproduct of tens column
68,475   Product

Notice that the subproduct of the units column in both methods of working the problem is 4565. The same is true for the subproduct of the tens column, 63,910. The zero is left off the subproduct in the second method because its only purpose is to hold a place. As long as the other figures are in their proper places, the zero is unnecessary.

Checking the Product

The common method of checking the accuracy of the product is to invert (turn over) the multiplicand with the multiplier, and work the problem again.

Multiplication		Check	
549	Multiplicand	63	Multiplicand
× 63	Multiplier	× 549	Multiplier
1647	Subproduct	567	Subproduct
3294	Subproduct	252	Subproduct
34,587	Product	315	Subproduct
		34,587	Product

If the problem is worked accurately in both instances, the products will be the same. If two different products are obtained, invert the problem back to its original form and try again.

Tips to Follow

A few tips are offered here to help provide speed and accuracy to multiplication work.

Be neat. This is very important. The customary method of multiplying eliminates end zeros (zeros which appear at the end of a number), therefore one must be very careful in writing each number in its proper place.

Be careful with carryover numbers. Remember that carryover numbers are added to the product of the two numbers being multiplied. The carryover numbers are not multiplied. The following example shows this. (Note: It is not necessary to write the carryover numbers as shown below.)

$$\begin{array}{r} ④ ⑤ \\ 846 \\ \times \quad 9 \\ \hline 7614 \end{array}$$

The first numbers to multiply are $9 \times 6 = 54$. Write the 4 in the units column beneath the bar and carry the 5 over to the tens column. Next, $9 \times 4 = 36$. To this is added the carryover number 5. $36 + 5 = 41$. Write the number 1 in the tens column beneath the bar and carry the 4 to the hundreds column. $9 \times 8 = 72$. Add the carryover number 4. $72 + 4 = 76$. Write the 6 in the hundreds column beneath the bar and the 7 in the thousands column beneath the bar. $9 \times 846 = 7614$.

When the multiplier consists of two or more numerals, be careful when adding the carryover numbers.

To quickly determine the product when multiplying by 10, 100, 1000, etc., add the correct number of zeros to the multiplicand. Example: $10 \times 218 = 2180$. Since there is one zero in 10, add one zero to 218 to obtain the product.

Use units in the product when they are used in the multiplicand. A unit was explained as a single quantity of like things. Most multiplication problems outside of the classroom involve some sort of units. If a small pork processing plant has 35 hogs in 27 separate pens, how many hogs does it have? $35 \times 27 = 945$. The product is not simply 945, but 945 hogs. The type of unit does not have to be written next to the multiplicand when working the problem, but remember what type of units are being multiplied, so the result will be a certain number of those units.

ACHIEVEMENT REVIEW
MULTIPLICATION

Find the answer to the following multiplication problems. If units are indicated, write the unit in the answer. Check your work.

1. $\begin{array}{r}146\\ \times \ 3\end{array}$	2. $\begin{array}{r}541\\ \times \ 4\end{array}$	3. $\begin{array}{r}4762\\ \times \ 78\end{array}$	4. $\begin{array}{r}3656\\ \times \ 29\end{array}$	5. $\begin{array}{r}723\\ \times \ 81\end{array}$
6. $\begin{array}{r}478\\ \times 212\end{array}$	7. $\begin{array}{r}5630\\ \times \ 385\end{array}$	8. $\begin{array}{r}6234\\ \times 3250\end{array}$	9. $\begin{array}{r}8768\\ \times 2345\end{array}$	10. $\begin{array}{r}\$251.45\\ \times \quad 42\end{array}$

11. $4,361.22
    × 46

12. $293.82
    × 45

13. $6,863.18
    × 34

14. $8,225.32
    × 2.85

15. $7,421.92
    × 1.67

16. A side of beef weighs 323 pounds and costs $0.98 per pound. How much does the side cost?

17. If a foresaddle of veal weighs 46 pounds and costs $2.75 per pound, what is the total cost?

18. Jim's catering truck gets 13 miles to every gallon of gas. How many miles can it travel on 78 gallons of gas?

19. When preparing meat loaf, it is required that 4 pounds of hamburger goes into each loaf. How many pounds of hamburger must be ordered when preparing 256 loaves?

20. A restaurant orders 28 beef rounds. Each round must average 45 pounds. How many pounds are ordered?

21. A restaurant orders 14 ribs of beef. Each rib must average 22 pounds. How many pounds are ordered?

22. Refer to problem 21. If the ribs of beef cost $1.88 per pound, what is the total cost of the order?

23. A wedding reception is catered for 575 people. The caterer charges $5.35 per person. What is the total bill?

24. A cook earns $65.00 a day. How much is earned in a year if the cook works 321 days?

25. A pastry shop receives 80 tins of frozen cherries. Each tin weighs 30 pounds. How many pounds are ordered?

DIVISION

Division is basically the method of finding out how many times one number is contained in another number. In this way, it is the reverse of multiplication.

Multiplication	$4 \times 9 = 36$
Division	$36 \div 9 = 4$
	$36 \div 4 = 9$

Look at division from another viewpoint. Assume that your employer promises to pay you $4.00 per hour. After working nine hours, you receive a check for only $27.00. Something is not right.

$$27 \div 9 = 3$$

You are only being paid $3.00 an hour. Using division can thus help you find out that a mistake was made.

Division is frequently used in the food service business, as shown in figure 2-5. As another example, a can contains 32 ounces of applesauce. How many guests can be served if each guest receives a 4-ounce portion?

$$32 \div 4 = 8$$

Fig. 2-5 Division is the method of operation used when dividing roll dough into units.

Eight guests can be served from the 32-ounce can of applesauce.

In division, the number to be divided is called the *dividend*. In the problem 70 ÷ 5 = 14, the number 70 is the dividend. The name for the number by which the dividend is divided (number 5 in this example), is the *divisor*. The result of dividing the dividend by the divisor is called the *quotient*. The quotient in this example is the number 14.

Sometimes the divisor is not contained in the dividend an exact amount of times. For example, 72 ÷ 5 = 14, with 2 left over because 5 is too large to be contained in 2. The number left over is called the *remainder*.

There are several division signs that can be used. Up to this point, division has been indicated by a bar with a dot above and below it (÷). This symbol is mainly used when the problem is simple, or when first stating a problem that is to be solved. Simple problems are also written using the fraction bar. For example, 72 ÷ 6 = 12 can also be written $^{72}\!/\!_6$ = 12.

For long division problems, use the division sign which looks like a closed parentheses sign with a straight line coming out of the top of it: $\overline{)\qquad}$.

Step-by-Step Division

In most reviews, division is discussed last because it involves both multiplication and subtraction (including carryover numbers and borrowing). Example:

$$7296 \div 16 = \,?$$

It is more convenient to write this problem in long division form.

thousands
hundreds
tens
units

Quotient goes here

Divisor $16\overline{)7\,2\,9\,6}$ Dividend

A long division problem is started from the left, rather than from the right as in addition, subtraction, and multiplication. The first thing to do is estimate how many times 16 is contained in 72 (actually 7200). It is known that 2 × 16 = 32. Twice that is 64. So it is estimated that 16 goes into 72 (7200) four (actually 400) times. The 4 is written above the bar in the hundreds place.

thousands
hundreds
tens
units

4   Start of quotient

Divisor $16\overline{)7\,2\,9\,6}$ Dividend

To make sure 4 is the correct figure, it is multiplied by 16. The product is 64. The 6 is written in the thousands column, under the 7, and the 4 in the hundreds column, under the 2. Then 64 is subtracted from 72. If the difference is less than 16, the estimate of 4 in the hundreds place of the quotient is correct. 72 − 64 = 8. Since 8 is less than 16, the estimate is correct.

thousands
hundreds
tens
units

$$\begin{array}{r} 4 \\ \text{Divisor } 16\overline{)7\,2\,9\,6}} \\ 6\,4 \\ \hline 8 \end{array}$$

Dividend

Product of 4 × 16

Difference of 72 − 64

The next step is to bring the 9 in the tens place of the dividend down to the right of the

8, giving 89. How many times is 16 contained in 89? Since $4 \times 16 = 64$, it is estimated that 89 contains 16 at least 5 times. $5 \times 16 = 80$. $89 - 80 = 9$, which is smaller than 16. (Note: Sometimes the estimate is too low or too high. When this happens, the estimate must be changed accordingly. If it is estimated that 56 contains 17 four times, the estimate is shown to be too high: $4 \times 17 = 68$, which is larger than 56.)

The problem, so far, looks like this:

```
                thousands
                hundreds
                tens
                units
            4 5
Divisor 16)7 2 9 6      Dividend
           6 4
            8 9
            8 0      Product of 5 × 16
              9      Difference of 89 − 80
```

Next, the 6 is brought down to the right of the 9 (left over from subtracting 80 from 89). How many times does 96 contain 16? It is known that $5 \times 16 = 80$. However, 6×16 equals exactly 96. The 96 is written beneath the 96 already there. The finished problem looks like this:

```
                thousands
                hundreds
                tens
                units
            4 5 6      Quotient
Divisor 16)7 2 9 6      Dividend
           6 4
            8 9
            8 0
              9 6
              9 6      Product of 6 × 16
```

The Remainder

The remainder in division is the number left over when the dividend does not contain the divisor exactly.

Example:

```
              31     Quotient
Divisor 37)1182     Dividend
           111      Product of 3 × 37
            72
            37      Product of 1 × 37
            35      Remainder
```

At this point, there are no more numerals to bring down from the dividend. There are no 37s contained in the number 35. Therefore, the leftover 35 is called the remainder. The remainder can also be written in fractional form.

$$1182 \div 37 = 31\frac{35}{37}$$

Checking the Division

The common method of checking division is to multiply the quotient by the divisor. The product of multiplying the divisor and the quotient together should be the same as the dividend.

Division

```
              135     Quotient
Divisor 21)2835     Dividend
           21
           73
           63
          105
          105
```

Check

$$
\begin{array}{rl}
135 & \text{Quotient} \\
\underline{\times\ 21} & \text{Divisor} \\
135 & \\
\underline{270} & \\
2835 & \text{Dividend}
\end{array}
$$

When the quotient includes a remainder, multiply the quotient and divisor together as before, then add the remainder to the product.

Check

$$
\begin{array}{rl}
212 & \text{Quotient} \\
\underline{\times\ 41} & \text{Divisor} \\
212 & \\
\underline{848} & \\
8692 & \text{Product} \\
\underline{+\ \ 6} & \text{Remainder added} \\
8698 & \text{Dividend}
\end{array}
$$

Division

$$
\begin{array}{rl}
& 212 \qquad \text{Quotient} \\
\text{Divisor}\quad 41\overline{)8698} & \quad \text{Dividend} \\
\underline{82} & \\
49 & \\
\underline{41} & \\
88 & \\
\underline{82} & \\
6 & \quad \text{Remainder}
\end{array}
$$

ACHIEVEMENT REVIEW
DIVISION

Find the quotient for the following division problems. Carry answers to three places to the right of the decimal point. Check all work.

1. $6\overline{)782}$
2. $4\overline{)973}$
3. $7\overline{)545}$
4. $26\overline{)858}$
5. $45\overline{)798}$

6. $47\overline{)2250}$
7. $123\overline{)8650}$
8. $254\overline{)10,250}$
9. $375\overline{)15,670}$
10. $59\overline{)890.75}$

11. $67\overline{)962.45}$
12. $29\overline{)648.28}$
13. $222\overline{)474.20}$
14. $355\overline{)555.42}$
15. $452\overline{)975.35}$

16. If a 14-pound leg of veal costs $51.08, what is the cost per pound?

17. A 340-pound side of beef is purchased costing $478.80. What is the cost per pound?

18. In one week, the Gourmet Catering Company's delivery truck traveled 385

miles. They used 35 gallons of gasoline. How many miles did they average per gallon?

19. A restaurant purchases 378 pounds of beef ribs. Each rib weighs 21 pounds. How many ribs are contained in the shipment?

20. A restaurant purchases 312 pounds of beef sirloin. The shipment contains 23 sirloins. What is the weight of each sirloin?

21. The Manor Catering Company's delivery truck averages 15 miles per gallon of gasoline. During thirty days of operation, the truck travels 3225 miles. How much gasoline is used?

22. A 49-pound beef round costs $83.21. What is the cost per pound?

23. A restaurant orders 312 pounds of pork loin. Each loin weighs 13 pounds. How many loins are contained in the shipment?

24. The cook at the Gallery Restaurant works 6 days a week and earns $292. per week. How much is earned in one day?

25. The chef at the Skylark Restaurant earns $25,925 per year. If this chef works 49 weeks per year, how much is earned in one week?

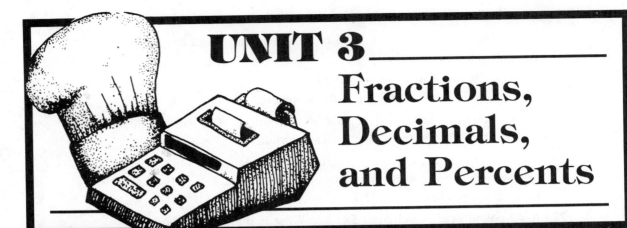

UNIT 3
Fractions, Decimals, and Percents

FRACTIONS

A *fraction* indicates one or more equal parts of a unit. For example, a cake is usually divided into 8 equal pieces, figure 3-1. If this is done, the following statements are true about the parts of the cake:

one part is ⅛ of the cake,
three parts are ⅜ of the cake
seven parts are ⅞ of the cake
eight parts are ⅛ of the cake or the whole cake.

Since fractions indicate the division of a whole unit into equal parts, the numeral placed above the division bar indicates the number of fractional units taken and is called the *numerator*. The numeral below the bar represents the number of equal parts into which the unit is divided and is called the *denominator*. Thus if an apple is cut into eight equal wedges, but only three of those wedges are used in garnishing a salad, the wedges used are represented by the fraction ⅜.

A *common fraction* is written with a whole number above the division bar and a whole number below the bar.

Fig. 3-1 An excellent example of a fractional part is shown when cutting a cake. Usually ⅛ is a serving portion.

A *proper fraction* is a fraction whose numerator is smaller than its denominator.

Example: $\dfrac{5}{8}\,\begin{matrix}\text{Numerator}\\\text{Denominator}\end{matrix}$

This type of fraction is in lowest possible terms when the numerator and denominator contain no common factor. (A *factor* refers to two or more numerals which, when multiplied together, yield a given product. Example: 2 and 4 are factors of 8.) The fraction ⅝ is in lowest possible terms because there is no common number by which both can be divided. (See simplification of fractions.)

An *improper fraction* is a fraction whose numerator is larger than its denominator and whose value is greater than a whole unit. If, for instance, 1¼ hams are expressed as an improper fraction, it would be expressed as ⁵⁄₄, since the one whole ham would be ⁴⁄₄, and the extra ¼ would make it ⁵⁄₄. Such fractions can be expressed as a mixed number by dividing the numerator by the denominator.

$$\frac{5}{4} = 1\frac{1}{4} \text{ mixed number}$$

A *mixed number* is a whole number mixed with a fractional part.

Examples: $1\frac{1}{2}, 4\frac{3}{4}, 6\frac{2}{3}$

Simplification of Fractions

Simplification is a method used to express a fraction in lower terms, without changing the value of the fraction. This is achieved by dividing the numerator and denominator of a fraction by the greatest factor (number) common to both. Examples:

$$\frac{16 \, (\div 8 \text{ greatest factor})}{24 \, (\div 8 \text{ greatest factor})} = \frac{2}{3}$$

$$\frac{12 \, (\div 4 \text{ greatest factor})}{16 \, (\div 4 \text{ greatest factor})} = \frac{3}{4}$$

The value of these fractions is not changed, but they have been simplified to their lowest terms.

A mixed number is usually expressed as an improper fraction when it is to be multiplied by another mixed number, a whole number, or a fraction. The first step is to express the mixed number as an improper fraction. This is done by multiplying the whole number by the denominator of the fraction, and then adding the numerator to the result. The sum is written over the denominator of the fraction.

Example:

$$2\frac{1}{4} \times 3\frac{2}{3} = \frac{9}{4} \times \frac{11}{3} = \frac{99}{12} = 8\frac{1}{4}$$

The whole number (2) is multiplied by the denominator of the fraction (4). To this result (8) the numerator (1) is added. The sum (9) is written over the denominator (4) creating the improper fraction ⁹⁄₄. The same procedure is followed in expressing the mixed number 3⅔ as the improper fraction ¹¹⁄₃. When the two mixed numbers are expressed as improper fractions, the product is found by multiplying the two numerators together and the two denominators together (⁹⁹⁄₁₂) and simplifying this improper fraction to lowest terms (8¼).

Adding and Subtracting Fractions

The addition and subtraction of fractions must be mastered to become successful in food preparation. Recipes are constantly being increased or decreased and many ingredients, such as herbs and spices, appear in fractional quantities. Another example of the use of fractions in food service is illustrated in figure 3-2.

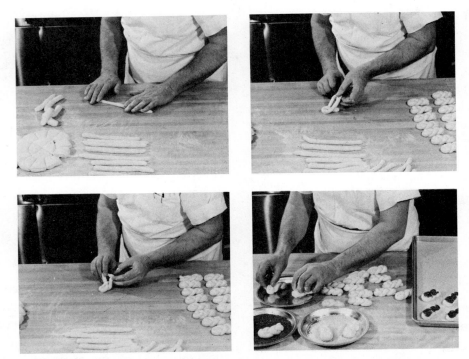

Fig. 3-2 A knowledge of fractions is important when dividing roll dough into units.

Before fractions can be added or subtracted they must have the same denominator. *Like fractions* are fractions that have the same denominator. To add or subtract like fractions, add or subtract the numerators and write the result over the common denominator. Examples of adding and subtracting like fractions:

$$\frac{2}{7} + \frac{1}{7} = \frac{3}{7}$$

$$\frac{6}{7} - \frac{5}{7} = \frac{1}{7}$$

Unlike fractions have different denominators. They are more difficult because only like things can be added or subtracted. Therefore, to add or subtract fractions that have unlike denominators, the fractions must first be expressed so the denominators are the same. To find this common denominator, multiply the two denominators together. The product will, of course, be common to both.

Example:

$$\frac{2}{3} + \frac{3}{4} = \frac{}{12} \quad \text{Denominator common to both fractions}$$

When a number is found that is a multiple of both denominators, the fractions are then expressed in terms of the common denominator, so ⅔ is ⁸⁄₁₂ and ¾ is ⁹⁄₁₂. These are now like fractions that can be added or subtracted without too much difficulty.

Add

$$\frac{2}{3} = \frac{8}{12}$$

$$+\frac{3}{4} = \frac{9}{12}$$

$$\frac{17}{12} = 1\frac{5}{12} \quad \text{Sum}$$

Subtract

$$\frac{3}{4} = \frac{9}{12}$$

$$-\frac{2}{3} = \frac{8}{12}$$

$$\frac{1}{12} \quad \text{Difference}$$

Note: In adding and subtracting unlike fractions, the common denominator may be any number that is a multiple of the original denominators, however, always use the least common denominator to simplify the work. The least common denominator is the smallest number that is a multiple of both denominators.

Example: If ¼ and ⅙ are to be added, the least common denominator is 12, since it is the smallest multiple of 4 and 6.

$$\frac{1}{4} = \frac{3}{12}$$

$$+\frac{1}{6} = \frac{2}{12}$$

$$\frac{5}{12}$$

Multiplying Fractions

Multiplying fractions is considered the simplest operation with fractions. When multiplying two fractions, multiply the two numerators and place the result over the result obtained by multiplying the two denominators.

Example: $\dfrac{5}{6} \times \dfrac{2}{3} = \dfrac{10}{18} = \dfrac{5}{9}$

If multiplying a whole number by a fraction, multiply the whole number by the numerator of the fraction, place the result over the denominator of the fraction, and divide the new numerator by the denominator.

Example: $\dfrac{16}{1} \times \dfrac{2}{3} = \dfrac{32}{3} = 10\dfrac{2}{3}$

Sometimes it is possible to simplify the problem before multiplying. In the example below, 5 is a factor of 15, because 15 contains 5 exactly 3 times.

Example: $\dfrac{\overset{3}{\cancel{15}}}{1} \times \dfrac{4}{\underset{1}{\cancel{5}}} = 12$

If the numerator and denominator can be divided evenly by the same number, simplify to lowest terms.

Example: $\dfrac{36}{54}$ numerator and denominator

can be divided evenly by 18,

resulting in $\dfrac{2}{3}$.

If multiplying by one or two mixed numbers, express the mixed number or numbers as improper fractions and proceed to multiply as with two fractions.

Examples: $\dfrac{6}{1} \times 2\dfrac{3}{8} = \dfrac{\overset{3}{\cancel{6}}}{1} \times \dfrac{19}{\underset{4}{\cancel{8}}} = \dfrac{57}{4} = 14\dfrac{1}{4}$

$2\dfrac{1}{3} \times 5\dfrac{3}{8} = \dfrac{7}{3} \times \dfrac{43}{8} = \dfrac{301}{24} = 12\dfrac{13}{24}$

Dividing Fractions

The operation of dividing fractions is perhaps the most difficult because it is necessary to invert (turn over) the divisor. Be careful to invert the correct fraction. After inverting the divisor, proceed to multiply the fractions.

Example A:

$$\frac{7}{8} \div \frac{1}{2} = \frac{7}{\overset{\cancel{8}}{4}} \times \frac{\overset{1}{\cancel{2}}}{1} = \frac{7}{4} = 1\frac{3}{4}$$

Step 1.   The divisor $\frac{1}{2}$ is inverted to $\frac{2}{1}$.

Step 2.   Multiply $\frac{7}{\overset{\cancel{8}}{4}} \times \frac{\overset{1}{\cancel{2}}}{1} = \frac{7}{4}$

Step 3.   The result, $\frac{7}{4}$, is an improper fraction. Expressed as a mixed number, it is 1¾.

Example B:

$$\frac{12}{1} \div \frac{1}{2} = \frac{12}{1} \times \frac{2}{1} = 24$$

Step 1.   The divisor $\frac{1}{2}$ is inverted to $\frac{2}{1}$.

Step 2.   Multiply $\frac{12}{1} \times \frac{2}{1} = 24$

ACHIEVEMENT REVIEW
FRACTIONS

A. Find the sum in each of the following addition problems. Simplify all answers to lowest terms.

1. $\frac{3}{7}$ $+\frac{2}{7}$

2. $\frac{5}{8}$ $+\frac{1}{8}$

3. $\frac{4}{9}$ $+\frac{2}{9}$

4. $\frac{7}{9}$ $+\frac{1}{9}$

5. $\frac{5}{16}$ $+\frac{7}{16}$

6. $\frac{1}{2}$ $+\frac{1}{8}$

7. $\frac{1}{16}$ $+\frac{1}{8}$

8. $\frac{7}{16}$ $+\frac{1}{4}$

9. $\frac{3}{4}$ $+\frac{1}{8}$

10. $1\frac{1}{8}$ $+1\frac{1}{4}$

11. $1\frac{3}{16}$ $+3\frac{3}{4}$

12. $\frac{23}{32}$ $\frac{15}{16}$ $+3\frac{1}{8}$

13. $2\frac{1}{2}$ $3\frac{3}{4}$ $+6\frac{1}{8}$

14. 15 $\frac{5}{12}$ $+\frac{7}{24}$

15. $\frac{1}{5}$ $\frac{8}{15}$ $+\frac{7}{15}$

30

16. $\dfrac{13}{50}$ $\dfrac{7}{10}$ $+\ \dfrac{4}{5}$

17. $\dfrac{1}{3}$ $\dfrac{5}{12}$ $+\ \dfrac{3}{8}$

18. $\dfrac{4}{9}$ $\dfrac{5}{12}$ $+\ \dfrac{5}{18}$

19. $\dfrac{1}{2}$ $\dfrac{1}{3}$ $+\ \dfrac{1}{5}$

20. $\dfrac{1}{4}$ $\dfrac{7}{12}$ $\dfrac{3}{16}$ $+\ 6\dfrac{5}{16}$

21. $6\dfrac{5}{6}$ $10\dfrac{5}{12}$ $+\ 13\dfrac{2}{3}$

22. $2\dfrac{5}{16}$ $6\dfrac{5}{8}$ $+\ 8\dfrac{1}{2}$

23. $20\dfrac{4}{5}$ $16\dfrac{3}{15}$ $+\ 18\dfrac{3}{10}$

24. $12\dfrac{1}{6}$ $9\dfrac{1}{3}$ $+\ 7\dfrac{1}{9}$

25. $12\dfrac{3}{8}$ $6\dfrac{1}{4}$ $+\ 10\dfrac{1}{2}$

B. Find the difference in each of the following subtraction problems. Simplify all answers to lowest terms.

1. $\dfrac{3}{7}$ $-\ \dfrac{2}{7}$

2. $\dfrac{5}{8}$ $-\ \dfrac{1}{8}$

3. $\dfrac{4}{9}$ $-\ \dfrac{2}{9}$

4. $\dfrac{7}{9}$ $-\ \dfrac{1}{9}$

5. $\dfrac{7}{16}$ $-\ \dfrac{5}{16}$

6. $\dfrac{1}{2}$ $-\ \dfrac{1}{8}$

7. $\dfrac{3}{4}$ $-\ \dfrac{2}{3}$

8. $\dfrac{1}{8}$ $-\ \dfrac{1}{16}$

9. $\dfrac{7}{16}$ $-\ \dfrac{1}{4}$

10. $3\dfrac{3}{4}$ $-\ 1\dfrac{3}{16}$

11. $9\dfrac{2}{3}$ $-\ 3\dfrac{1}{3}$

12. $12\dfrac{1}{4}$ $-\ 5\dfrac{5}{16}$

13. $15\dfrac{3}{4}$ $-\ 12\dfrac{2}{3}$

14. $14\dfrac{7}{24}$ $-\ 6\dfrac{5}{16}$

15. $45\dfrac{5}{8}$ $-\ 32\dfrac{7}{16}$

16. $22\dfrac{2}{3}$ $-\ 12\dfrac{5}{6}$

17. $18\dfrac{7}{16}$ $-\ 13\dfrac{1}{4}$

18. $16\dfrac{1}{3}$ $-\ 9\dfrac{2}{5}$

19. $23\dfrac{5}{18}$ $-\ 7\dfrac{5}{12}$

20. $42\dfrac{23}{32}$ $-\ 23\dfrac{15}{16}$

21. $32\dfrac{5}{8}$ 22. $21\dfrac{3}{16}$ 23. $15\dfrac{5}{9}$ 24. $20\dfrac{5}{12}$ 25. $3\dfrac{4}{5}$

$-18\dfrac{1}{4}$ $-16\dfrac{1}{8}$ $-8\dfrac{2}{3}$ $-12\dfrac{3}{16}$ $-7\dfrac{3}{15}$

C. Find the product in each of the following multiplication problems.

1. $\dfrac{5}{8} \times \dfrac{1}{2} =$

2. $\dfrac{5}{16} \times \dfrac{1}{5} =$

3. $\dfrac{7}{8} \times \dfrac{1}{4} =$

4. $1\dfrac{3}{4} \times 2\dfrac{3}{8} =$

5. $5\dfrac{1}{2} \times 3\dfrac{1}{2} =$

6. $4\dfrac{1}{4} \times 6\dfrac{1}{2} =$

7. $1\dfrac{3}{4} \times 4\dfrac{3}{8} =$

8. $36 \times 5\dfrac{3}{4} =$

9. $45 \times 7\dfrac{1}{2} =$

10. $1\dfrac{1}{8} \times 1\dfrac{1}{4} =$

11. $4\dfrac{3}{4} \times 5\dfrac{1}{2} =$

12. $32 \times 7\dfrac{1}{2} =$

13. $18 \times 9\dfrac{5}{9} =$

14. $\dfrac{2}{3} \times 12\dfrac{1}{3} =$

15. $\dfrac{5}{8} \times 7\dfrac{3}{8} =$

16. $24\dfrac{1}{3} \times 7\dfrac{2}{3} =$

17. $22\dfrac{5}{16} \times 4 =$

18. $7\dfrac{1}{4} \times 3 =$

19. $8\dfrac{2}{9} \times 3\dfrac{1}{2} =$

20. $8\dfrac{2}{5} \times 3\dfrac{3}{5} =$

21. $8\dfrac{2}{3} \times 2\dfrac{1}{4} =$

22. $10\dfrac{3}{4} \times 3\dfrac{1}{2} =$

23. $22 \times 5\dfrac{1}{4} =$

24. $26 \times 4\dfrac{1}{2} =$

25. $3\dfrac{5}{8} \times \dfrac{1}{2} =$

D. Find the quotient in each of the following division problems.

1. $\dfrac{15}{16} \div \dfrac{3}{4} =$

2. $\dfrac{1}{2} \div \dfrac{1}{4} =$

3. $\dfrac{15}{16} \div 2 =$

4. $\dfrac{1}{4} \div \dfrac{7}{8} =$

5. $\dfrac{3}{4} \div \dfrac{3}{16} =$

6. $\dfrac{3}{8} \div 4 =$

7. $\dfrac{7}{16} \div \dfrac{3}{16} =$

8. $1\dfrac{1}{4} \div \dfrac{3}{8} =$

9. $1\dfrac{1}{2} \div 6 =$

10. $\dfrac{4}{5} \div \dfrac{2}{3} =$

11. $2\dfrac{5}{8} \div 7 =$

12. $10 \div \dfrac{1}{4} =$

13. $\dfrac{1}{4} \div 10 =$

14. $\dfrac{3}{4} \div 2 =$

15. $\dfrac{5}{8} \div \dfrac{1}{2} =$

16. $\dfrac{5}{16} \div 3 =$

17. $4\dfrac{3}{4} \div \dfrac{1}{2} =$

18. $2\dfrac{1}{4} \div 1\dfrac{1}{2} =$

19. $6\dfrac{1}{8} \div 4\dfrac{1}{4} =$

20. $7\dfrac{5}{8} \div 2\dfrac{1}{4} =$

21. $9\dfrac{3}{8} \div \dfrac{1}{4} =$

22. $10\dfrac{2}{3} \div \dfrac{1}{2} =$

23. $20\dfrac{1}{2} \div \dfrac{1}{4} =$

24. $\dfrac{7}{8} \div \dfrac{1}{4} =$

25. $12\dfrac{3}{4} \div \dfrac{1}{3} =$

DECIMALS

A decimal is based on the number ten. The decimal system refers to counting by tens and powers of ten. The term decimals refers to decimal fractions. *Decimal fractions* are those fractions which are expressed with denominators of 10 or powers of 10.

Examples: $\dfrac{1}{10}$ $\dfrac{9}{100}$ $\dfrac{89}{1000}$ $\dfrac{321}{10,000}$

Instead of writing a fraction, a point (.) called a *decimal point* is used to indicate a decimal fraction.

Examples: $\dfrac{1}{10} = 0.1$ $\dfrac{9}{100} = 0.09$

$\dfrac{89}{1000} = 0.089$ $\dfrac{321}{10,000} = 0.0321$

Numbers go in both directions from the decimal point. The place value of the numbers to the left starts with the units or ones column, and each column (moving left) is an increasing multiple of 10.

Thousands	Hundreds	Tens	Units or Ones
1000	100	10	1.

To the right of the decimal point, each column is one-tenth of the number in the column immediately to its left. For example, one-tenth of one is $\frac{1}{10}$. Thus the decimals to the right of the decimal point are 0.1, 0.01, 0.001, 0.0001, and so on. These numbers stated as decimal fractions are $\frac{1}{10}$, $\frac{1}{100}$, $\frac{1}{1000}$, and $\frac{1}{10,000}$.

Decimal fractions differ from common fractions because they have 10 or a power of 10 for the denominator, whereas common fractions can have any number for the denominator. To simplify writing a decimal fraction, the decimal point is used.

Example: Express the decimal fraction $^{725}/_{1000}$ as its equivalent using a decimal point.

1. To convert the decimal fraction to a decimal, first write the numerator (725).

2. Count the number of zeros in the denominator and place the decimal point according to the number of zeros. There must always be as many decimal places as there are zeros in the denominator. (0.725)

Often, when writing a decimal fraction as a decimal, it is necessary to add zeros to the left of the numerator before placing the decimal point, to indicate the value of the denominator.

Example: $^{725}/_{10,000} = 0.0725$ Read as seven hundred twenty-five ten thousandths.

When a number is made up of a whole number and a decimal fraction, it is referred to as a *mixed decimal fraction*. To write a mixed decimal fraction, the whole number is written to the left of the decimal point and the fractional part to the right of the decimal point. Example: $7^{135}/_{1000} = 7.135$. The decimal point is read as "and," so to read this mixed decimal fraction, the whole number is read first, then the decimal point as "and." Next, read the fraction as a whole number and state the denominator. Following this procedure, 7.135 is read seven and one hundred thirty-five thousandths.

To add or subtract decimal fractions, keep all whole numbers in their proper column and all decimal fractions in their proper column. Remember that the decimal point separates whole numbers from fractional parts. It is therefore very important that decimal points are directly in line with each other.

Examples:

Adding Decimal Fractions

         2.135
         7.43
         4.008
      ___1.125___   Note: The decimal point in the
        14.698   sum goes under the decimal point
                    of the other numbers.

Subtracting Decimal Fractions

      9.825 Minuend
    − 5.450 Subtrahend
      _____
      4.375 Difference Note: The decimal point in
                          the difference goes under
                          the decimal point in the
                          minuend and subtrahend.

To multiply decimal fractions, follow the same procedure as when multiplying whole numbers to find the product. To locate the decimal point in the product, count the number of decimal places in both the multiplicand and the multiplier. The number of decimal places counted in the product is equal to the sum of those in the multiplicand and multiplier.

Example:

         4.32   Multiplicand
      × 0.06   Multiplier
      _____
       .2592   Product

There are four decimal places in the multiplicand and multiplier. Therefore, four decimal places are counted from right to left in the product.

In many cases, the total number of decimal places in the multiplicand and multiplier exceeds the number of numerals that appear in the product.

In such cases, *ciphers* (zeros) are added to the left of the digits in the product to complete the decimal places needed.

Example:

         .445   Multiplicand
      × .16   Multiplier
      _____
        2670
        445
      _____
      .07120   Product

Note: A cipher is added to the product to complete the 5 decimal places required.

To divide decimal fractions, proceed as if the numbers were whole numbers and place the decimal point as follows:

1. When dividing by whole numbers, place the decimal point in the answer directly above the decimal point in the dividend.

Examples:

          .06   Use zeros as needed in the
      6)0.36   quotient to hold a place.
        36
        ___

          .8   Zeros are not needed in the
      5)4.0   quotient to hold a place.
        4.0
        ___

2. When dividing a whole number or mixed decimal by a mixed decimal or decimal fraction, change the divisor and dividend so the divisor becomes a whole number. This is accomplished by multiplying both the dividend and divisor by the same power of ten. The divisor and dividend can be multiplied by the same power of ten without changing the value of the division.

Example:

$$\begin{array}{r} 12. \quad \text{Quotient} \\ \text{Divisor} \quad .25\overline{)3.00} \quad \text{Dividend} \\ \underline{2\;5} \\ 50 \\ \underline{50} \end{array}$$

In this example, the divisor 0.25 is made into the whole number 25 by multiplying by 100,

moving the decimal point 2 places to the right. Since the dividend must also be multiplied by 100, the decimal point in the dividend is also moved two places to the right, so 3 becomes 300. The decimal point in the quotient is always placed directly over the decimal point in the dividend. The answer is 12, a whole number. Note: When moving a decimal point, an arrow is used to show where the decimal point is to be moved.

ACHIEVEMENT REVIEW
DECIMALS

A. Change the following fractions to decimals.

1. $\dfrac{9}{10}$ 6. $\dfrac{7}{10}$ 11. $\dfrac{28}{100}$

2. $\dfrac{15}{100}$ 7. $\dfrac{29}{100}$ 12. $\dfrac{843}{1000}$

3. $\dfrac{3}{10}$ 8. $\dfrac{861}{100,000}$ 13. $\dfrac{676}{10,000}$

4. $\dfrac{232}{1000}$ 9. $\dfrac{429}{1000}$ 14. $\dfrac{39}{1000}$

5. $\dfrac{451}{1000}$ 10. $\dfrac{785}{10,000}$ 15. $\dfrac{45}{10,000}$

B. Write the following decimals and mixed decimal fractions in words.

1. 0.6   6. 0.0578   11. 6.725
2. 0.29   7. 0.06254   12. 7.42
3. 0.859   8. 0.67187   13. 9.135
4. 0.18   9. 2.2   14. 3.41
5. 0.002   10. 6.3   15. 2.567

C. Write as decimals.

1. Five tenths
2. Fourteen hundredths

3. One hundred eight thousandths
4. Sixteen hundredths
5. Two thousands four hundred twelve ten thousandths
6. Sixty-eight ten thousandths
7. One hundred twenty-two thousandths
8. Four thousand one hundred twenty-three ten thousandths
9. Sixty-four hundred thousandths
10. Three and five tenths
11. Four and twenty-five hundredths
12. Sixteen hundred thousandths
13. One hundred thousandths
14. Eight and nine hundredths
15. Twenty-three and six hundred twenty-five ten thousands

D. Find the sum in each problem.

1. 0.46 + 0.062 + 8.169 + 0.046
2. 0.58 + 0.675 + 6.225 + 9.323
3. 0.015 + 0.702 + 10.318 + 12.962
4. 0.506 + 0.015 + 0.711 + 6.25 + 16.37
5. 0.046 + 0.002 + 643 + 6.1675

E. Find the difference in each problem.

1. 9.765 − 0.046
2. 8 − 0.123
3. 1 − 0.685
4. 0.0622 − 0.0421
5. 139.371 − 123.218

F. Find the product in each problem. Round the answers to the nearest ten thousandth when necessary.

1. 412 × 0.52
2. 0.922 × 1.52
3. 0.0524 × 0.132
4. 6.53 × 0.38
5. 7.323 × 5.452

G. Find the quotient in each problem. Give quotients to 3 decimal places when necessary.

1. 0.49 ÷ 7
2. 15 ÷ 4.8
3. 945 ÷ 500
4. 0.0684 ÷ 24
5. 46.76 ÷ 400
6. 4 ÷ 0.25
7. 2.65 ÷ 1.5
8. 6.5 ÷ 2.45
9. 45.5 ÷ 3.5
10. 54.25 ÷ 4.8

PERCENT

Percent (%) means of each hundred. Thus 5 percent means 5 out of every 100. This same 5 percent can also be written 0.05 in decimal form. In fraction form it is 5/100.

Percents are mainly used to express a rate of each hundred. If 50 percent of the customers in a restaurant select beef entrees, a rate of a whole is being expressed. The whole is represented by 100 percent. In this example, all of the customers that enter the restaurant equal 100 percent. The use of percents in expressing a rate is a common practice in the food service business. For example, food cost, labor cost, and profits are almost always expressed by percents. Another example of percent in food service is shown in figure 3-3.

To find what percent one number is of another number, divide the number that represents the part by the number that represents the whole. Example: A side of beef weighs 480 pounds. The forequarter weighs 220 pounds. What percent of the side is the forequarter?

The 480-pound amount represents the whole side of beef. The 220 pounds represents only a fractional part of the side. As a fraction, it is written $^{220}/_{480}$, simplified to lowest terms = $^{11}/_{24}$. To express a common fraction as a percent, the numerator is divided by the denominator. The division must be carried out two places (hundreds place) for the percent. Carrying the

Fig. 3-3 An example of percent. 10% of the hot cross buns have been removed from the sheet pan.

division three places behind the decimal gives the tenth of a percent, which makes the percent more accurate. The finished problem looks like this:

$$\frac{0.458}{480)\overline{220.000}} = 45.8\%$$

$$\begin{array}{r} 192\ 0 \\ \hline 28\ 00 \\ 24\ 00 \\ \hline 4\ 000 \\ 3\ 840 \\ \hline \end{array}$$

Percent means hundredths so the decimal is moved two places to the right when the percent sign is used.

When a percent is expressed as a common fraction, the given percent is the numerator and 100 is the denominator.

Examples:

$$40\% = \frac{40}{100} \text{ or } \frac{2}{5}$$

$$20\% = \frac{20}{100} \text{ or } \frac{1}{5}$$

When a percent is changed to a decimal fraction, the percent sign is removed and the decimal point is moved two places to the left.

Examples:
$$25.5\% = 0.255$$
$$85.6\% = 0.856$$

When a decimal fraction is expressed as a percent, move the decimal point two places to the right and place the percent sign to the right of the last figure.

Examples:
$$0.255 = 25.5\%$$
$$0.856 = 85.6\%$$

The above percents are read as twenty-five and five-tenths percent, and eighty-five and six-tenths percent.

Remember the following points when dealing with percents:

- When a number is compared with a number larger than itself, the result is always less than 100 percent. Example: 48 is 80% of 60 since $^{48}/_{60} = ^4/_5 = 80\%$.
- When a number is compared with itself, the result is always 100 percent. Example: 48 is 100% of 48 since $^{48}/_{48} = 1 = 100\%$.
- When a number is compared with a number smaller than itself, the result is always more than 100 percent. Example: 48 is 120% of 40 since $^{48}/_{40} = ^6/_5 = 120\%$.
- When taking a percent of a whole number the method of operations is to multiply.

Example: If a restaurant takes in $8,925.00 in one week, but only 26 percent of that amount was profit, what is the profit?

$$\begin{array}{r} \$8,925 \\ \times\ 0.26 \\ \hline 53550 \\ 17850 \\ \hline \$2,320.50 \end{array} \quad \text{Profit}$$

ACHIEVEMENT REVIEW
PERCENT

A. Express the following common fractions as percents.

1. $\dfrac{5}{8}$ 2. $\dfrac{7}{8}$ 3. $\dfrac{4}{5}$

4. $\dfrac{3}{8}$ 8. $\dfrac{2}{3}$ 12. $\dfrac{5}{9}$

5. $\dfrac{5}{16}$ 9. $\dfrac{8}{9}$ 13. $\dfrac{3}{4}$

6. $\dfrac{3}{16}$ 10. $\dfrac{5}{12}$ 14. $\dfrac{3}{10}$

7. $\dfrac{5}{6}$ 11. $\dfrac{4}{9}$ 15. $\dfrac{7}{10}$

B. Express the following as common fractions, whole numbers, or mixed numbers.

1. 5%   4. 100%   7. 60%   10. 20%   13. 125%
2. 50%   5. 75%   8. 65%   11. 36%   14. 200%
3. 25%   6. 28%   9. 24%   12. 48%   15. 140%

C. Solve the following. Round answers to nearest hundredth.

1. 25% of 300   9. 42½% of $562.00
2. 30% of 687   10. 65% of $1,250.00
3. 3% of 520   11. 75% of $650.00
4. 42% of 400   12. 32% of $465.90
5. 28% of 980   13. 28% of $2,250.00
6. 12½% of 80   14. 46% of $3,670.50
7. 35% of $38.00   15. 54% of $4,880.20
8. 38% of $465.50

16. A 643-pound side of beef is ordered. The rib cut weighs 24 pounds. What percent of the side was the rib?

17. A 550-pound side of beef is ordered. The chuck cut weighs 38 pounds and the round cut weighs 48 pounds. What percent of the side is the chuck? What percent is the round?

18. A party for 250 people is booked. The cost of the party is $560.00 If they are given a 12 percent discount on their total bill, what is the cost of the party?

19. A party for 100 people is booked. The cost of the party is $375.00. If they are given a 15 percent discount on their total bill, what is the cost of the party?

20. If a restaurant takes in $2,600.00 in one week and 37 percent of that amount is for expenses, how much are the expenses?

21. If a restaurant takes in $4,250.00 in one week and 32 percent of that amount is profit, how much is profit?

22. If a 45-pound beef round is roasted and 12 pounds are lost through shrinkage, what percent of the round is lost through shrinkage?

23. If a 22-pound rib of beef is roasted and 6 pounds are lost through shrinkage, what percent of the rib is lost through shrinkage?

24. If a restaurant's gross receipts for one week total $4,000.00 of which $1,500.00 is profit, what percent of the gross receipts is profit?

25. If a restaurant's gross receipts for one week total $3,800.00 of which $2,500.00 are expenses, what percent of the gross receipts are expenses?

26. Mr. Hayes purchased four new slicing machines at $1285.00 each. For buying in quantity, he is given a 4½ percent discount on the total bill. What is the amount he paid?

27. Mrs. Hill purchased three new coffee urns at $1375.00 each. For buying in quantity, she is given a 3½ percent discount on the total bill. What is the amount she paid?

28. Hasenour's Restaurant purchased a 23-pound rib of beef at $2.23 per pound. Through trimming and roasting, 20 percent of the original amount is lost. How much money is lost through shrinkage and trimming?

29. Mr. Bryce ordered the following supplies for his restaurant.

    6 — # 10 cans tomato puree   @ $2.24 per can
    6 — # 10 cans green beans   @ $2.65 per can
    6 — # 10 cans cherries   @ $3.40 per can
    6 — # 10 cans peaches   @ $2.80 per can
    6 — # 10 cans corn   @ $2.18 per can

    He is given a 2 percent discount on the total bill. What is the amount he paid?

30. Mrs. Flynn ordered the following supplies for her restaurant.

    6 — # 10 cans beets   @ $2.15 per can
    6 — # 10 cans pears   @ $3.28 per can
    6 — # 10 cans pineapple   @ $3.48 per can
    6 — # 10 cans carrots   @ $2.16 per can
    6 — # 10 cans bamboo shoots   @ $4.45 per can

She is given a 2½ percent discount on the total bill. What is the amount she paid?

31. A restaurant operator purchased various pieces of equipment directly from the manufacturer, amounting to a total of $2,550.00. The manufacturer gave a 4½ percent discount for buying direct. What is the total cost?

32. A restaurant operator purchased various pieces of equipment directly from the manufacturer, amounting to a total of $4,678.00. The manufacturer gave a 6½ percent discount for buying direct. What is the total cost?

33. If a restaurant's gross receipts for one week are $4,850.00 and 30 percent of that amount is for labor, 41 percent for food cost, and 12 percent for miscellaneous items, how much is the profit?

34. If a restaurant's gross receipts for one week are $6,580.00 and 34 percent of that amount is for labor, 39 percent for food cost, and 15 percent for miscellaneous items, how much is the profit?

35. If a restaurant takes in $6,850.00 in one week and the cost of food sold is $2,950, what percent is the food cost?

36. Haines Gourmet Restaurant purchased 6 sirloins of beef weighing a total of 114 pounds. Through boning and trimming 19 pounds are lost. What percent is lost through boning and trimming?

37. If a 35-pound round of beef is roasted and 13 pounds are lost through shrinkage, what percent of the round is lost through shrinkage?

38. If a restaurant's gross receipts for one week total $4,280.00 but only 23 percent of that amount is profit, how much are expenses?

PART TWO

Preparation Crew— Operational Procedures

In the past, the members of the preparation crew in a food service establishment were only concerned with doing the individual task they were hired to perform. Other details of the operation, such as controlling portions, food production reports, and even converting recipes were management's responsibility. Today this situation can no longer exist if an operation is to survive. Control in all areas of production has become everyone's job. This is the only way management can control the high cost of food and keep menu items at a competitive price.

There are a number of situations that occur in the preparation area of a commercial kitchen with which the preparation crew must be familiar if the operation is to run efficiently. These situations are daily occurrences that require the attention of each member of the crew.

The preparation crew usually consists of the chef, cooks, butcher, baker, and pantry or salad people. Every member of the crew should understand the following operational procedures contained in this section.

- Weights and measures
- Portion control
- Converting standard recipes
- Daily food production report
- Back-of-the-house business forms
- Formulas to remember

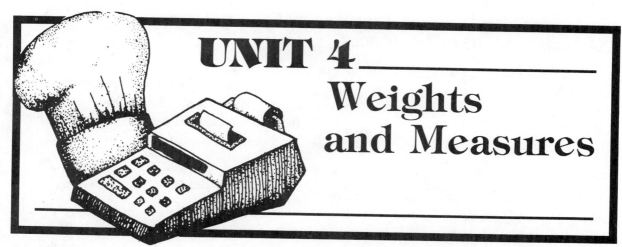

UNIT 4
Weights and Measures

Recipes or formulas used in a commercial kitchen are given in weights or measures. The more exact recipes or formulas are stated in weights. In this way, ingredient amounts can be found more accurately. When measuring ingredients one is not always exact; much depends on how firmly the ingredients are packed into the measuring device, and what the individual considers full. The common measures used are teaspoons, tablespoons, cups, pints, quarts, and gallons. These are usually abbreviated when stated in recipes. Figure 4-1 gives the common abbreviations used.

The relationship of the various measures and weights is given in figure 4-2. It can be used to convert from measures to weights or weights to measures. For example, if 2 pounds of liquid milk are required in a recipe, this can be measured as 1 fluid quart since 2 pounds of a liquid equals 1 fluid quart; or if 1 pound of whole eggs is needed, this can be measured as 1 fluid pint. The table in figure 4-2 can help save production time.

tsp.	teaspoon
tbsp.	tablespoon
pt.	pint
qt.	quart
gal.	gallon
oz.	ounce
lb.	pound
bch.	bunch
doz.	dozen
ea.	each
crt.	crate

Fig. 4-1 Abbreviations of measures used in recipes.

1 pinch	1/2 teaspoon (approx.)
3 teaspoons	1 tablespoon
16 tablespoons	1 cup
2 cups	1 pint
2 pints	1 quart
4 quarts	1 gallon
16 ounces	1 pound
8 ounces (liquid and eggs)	1 fluid cup
1 pound (liquid and eggs)	1 fluid pint
2 pounds (liquid and eggs)	1 fluid quart
8 pounds (liquid and eggs)	1 fluid gallon

Fig. 4-2 Equivalents of measures.

MEASURING AND WEIGHING DEVICES

There are many measuring and weighing instruments used in food service operations. Some of these are illustrated in figure 4-3, and described in the following paragraphs.

Scoops and Dippers

Scoops or dippers are used to serve many foods and also to control the portion size. The various sizes of scoops or dippers are designated by a number which appears on the lever that releases the item from the scoop. Figure 4-4 gives the relationship of the scoop number to the approximate capacity in ounces. It also gives the approximate content of each scoop in cups or tablespoons.

Ladles

Ladles are used to serve stews, soups, sauces, gravies, dressings (figure 4-5), cream dishes, and so on, when portion control and uniform servings are desired. Figure 4-6 shows the ladle sizes most frequently used.

The Baker's Scale

The *baker's scale* is the best type of scale to use because it ensures accuracy. This scale has a twin platform as shown in figure 4-7. On the platform to the left is placed a metal scoop, in which the food to be weighed is placed. On the platform to the right is placed a special weight equal to the weight of the scoop. A beam that is graduated in ¼ ounces runs horizontally across

Fig. 4-3 An assortment of measuring and weighing utensils used in food service operations.

Scoop or Dipper No.	Approximate Weight	Scoop or Dipper No.	Level Measure
8	5 oz.	8	1/2 cup
10	4 oz.	10	2/5 cup
12	3 oz.	12	1/3 cup
16	2 to 2 1/2 oz.	16	1/4 cup
20	1 2/3 oz.	20	3 1/5 tablespoons
24	1 1/2 oz.	24	2 2/3 tablespoons
30	1 1/4 oz.	30	2 1/5 tablespoons
40	1 oz.	40	1 3/5 tablespoons

Fig. 4-4 Scoop and dipper sizes and appropriate weights.

Fig. 4-5 Ladles are used when dishing up dressings to control portion size and achieve uniform servings.

1/4 cup	2 oz.
1/2 cup	4 oz.
3/4 cup	6 oz.
1 cup	8 oz.

Fig. 4-6 Ladle sizes.

the front of the scale. The beam has a weight attached to it, as shown in figure 4-7.

This weight is placed on the number of ounces one wishes to weigh. The beam is graduated in ¼ ounces up to 16 ounces (1 pound). If a larger amount is to be weighed, additional metal weights of 1, 2, and 4 pounds are provided. When using the baker's scale, always balance the scale before setting the weights for a given amount.

To weigh 8 ounces of whole eggs, for example, place a container large enough to hold the whole eggs on the left platform and balance the scale. Move the weight on the beam 8 ad-

Fig. 4-7 Twin platform baker's scale used in most bake shops to ensure accuracy when weighing ingredients.

ditional ounces. Add the whole eggs until the scale balances again.

If 12¾ ounces of flour are to be weighed, place the metal scoop on one platform and the special balancing weight on the other platform. This brings the two platforms to a complete balance. The weight on the scaling beam is set on 12¾ ounces. Flour is placed in the metal scoop, figure 4-8, until the two platforms balance a second time. For weighing ½ ounce of baking soda, a piece of paper can be placed on the left platform, and the platforms balanced. Set the balance weight at ½ ounce and add the baking soda until the platforms balance again. This scale can be used to weigh up to 10 pounds.

The student cook is expected to be able to use the baker's scale correctly, and should therefore practice weighing ingredients on this scale whenever possible. When practicing, it is best to use items such as flour, salt, sugar, and water.

Fig. 4-8 Student weighing flour on a baker's scale. Weights are more accurate than measures.

ACHIEVEMENT REVIEW
USING THE BAKER'S SCALE

Total the amounts of each ingredient in the 8 formulas given and write the total below each formula.

Example:

Ingredients	Pounds	Ounces
Sugar	1	8
Flour	4	6
Salt	–	1
Shortening	1	12
Total	6 lb.	27 oz. = 7 lb. 11 oz.

(Note: There are 16 ounces in a pound, so for each 16 ounces, carry one pound to the pound column. 27 ounces equals 1 pound and 11 ounces, so 6 pounds 27 ounces equals 7 pounds 11 ounces.)

Proceed to weigh each ingredient separately using sugar or rice to represent each ingredient. When this is done, weigh the total amount of all ingredients weighed. The amount should equal the sum of all the weights of ingredients listed.

Formula 1:

Ingredients	Pounds	Ounces
Shortening	1	6
Bread Flour	3	12
Salt	–	1
Sugar	2	10
Baking Powder	–	3
Total		

Formula 2:

Ingredients	Pounds	Ounces
Butter	–	12
Pastry Flour	3	8
Sugar	1	8
Salt	–	1
Dry Milk	–	3
Total		

Formula 3:

Ingredients	Pounds	Ounces
Shortening	1	8
Pastry Flour	2	10
Baking Powder	–	4
Baking Soda	–	½
Sugar	1	12
Salt	–	2 ¼
Total		

Formula 4:

Ingredients	Pounds	Ounces
Butter	–	14
Cake Flour	3	12
Baking Powder	–	3 ½
Salt	–	½
Sugar	2	10
Dry Milk	–	6
Total		

Formula 5:

Ingredients	Pounds	Ounces
Sugar	2	6
Baking Powder	–	¾
Salt	–	½
Flour	4	10
Cornstarch	–	5
Total		

Formula 6:

Ingredients	Pounds	Ounces
Shortening	2	6
Pastry Flour	4	8
Dry Milk	–	4
Bread Flour	1	2
Salt	–	½
Sugar	2	12
Total		

Formula 7:

Ingredients	Pounds	Ounces
Butter	1	12
Pastry Flour	4	6
Dry Milk	–	3 ½
Shortening	1	6
Sugar	2	10
Salt	–	¾
Total		

Formula 8:

Ingredients	Pounds	Ounces
Shortening	1	8
Bread Flour	3	12
Dry Milk	–	6
Sugar	2	12
Cornstarch	–	5
Salt	–	¾
Total		

PORTION SCALE

The *portion scale*, figure 4-9, is used for measuring food servings such as a 3-ounce serving of cooked roast beef, a 2½-ounce serving of cooked ham, or for portioning a 5-ounce hamburger steak or a 4½-ounce patty of sausage. Portioning food servings is an important part of any food service operation. If the exact cost of each item sold is to be determined it is necessary to know the yield of each food. For example, if 15 pounds of beef round costs $25.00, the cost per serving cannot be determined until a serving portion and yield are established after the meat has been roasted.

The portion scale shown in figure 4-10 is used to weigh quantities up to 32 ounces (2 pounds). Each number on the movable dial rep-

Fig. 4-10 Portion scale used to portion all kinds of food in the commercial kitchen.

resents 1 ounce. Each mark between the numbers represents ¼ ounce. For instance, the first mark past the pointer is ¼ ounce. The longer line next to it is ½ ounce. The short line next in order is ¾ ounce and the long line which comes next is 1 full ounce. The platform at the top of the scale is attached to a metal stem which fits into the scale. This platform can be removed for washing, but care must be taken when replacing it, so it fits properly and performs accurately. When the platform is properly placed, the pointer rests at 0, which represents 32 ounces. When weighing amounts that fit on the platform, first place waxed paper or freezer paper on the platform. Then, using the handle to move the scale dial, move the dial to the left until the pointer is again at zero. This action accounts for the weight of the paper. Place enough of the item being weighed on the paper until the pointer is exactly at the amount needed. Example: If por-

Fig. 4-9 Portioning meat on a portion scale helps control serving size and cost.

tioning a 6-ounce ground beef patty, place enough beef on the platform so that the pointer points directly at the 6, indicating that 6 ounces have been obtained. When weighing amounts that do not fit properly on the platform, use a light aluminum cake or pie pan to hold the item. Before

weighing the item, however, be sure to balance the pan, using the same method as for balancing the paper.

Some approximate weights and measures of common foods are listed in figure 4-11.

Food Product	Tbsp.	Cup	Pt.	Qt.
Allspice	1/4 oz.	4 oz.	8 oz.	1 lb.
Apples, Fresh, Diced	1/2 oz.	8 oz.	1 lb.	2 lb.
Bacon, Raw, Diced	1/2 oz.	8 oz.	1 lb.	2 lb.
Bacon, Cooked, Diced	2/3 oz.	10 1/2 oz.	1 lb. 5 oz.	2 lb. 12 oz.
Baking Powder	3/8 oz.	6 oz.	12 oz.	1 lb. 8 oz.
Baking Soda	3/8 oz.	6 oz.	12 oz.	1 lb. 8 oz.
Bananas, Sliced	1/2 oz.	8 oz.	1 lb.	2 lb.
Barley	—	8 oz.	1 lb.	2 lb.
Beef, Cooked, Diced	3/8 oz.	5 1/2 oz.	11 oz.	1 lb. 6 oz.
Beef, Raw, Ground	1/2 oz.	8 oz.	1 lb.	2 lb.
Bread Crumbs, Dry	1/4 oz.	4 oz.	9 oz.	1 lb. 2 oz.
Bread Crumbs, Fresh	1/8 oz.	2 oz.	4 oz.	8 oz.
Butter	1/2 oz.	8 oz.	1 lb.	2 lb.
Cabbage, Shredded	1/4 oz.	4 oz.	8 oz.	1 lb.
Carrots, Raw, Diced	5/16 oz.	5 oz.	10 oz.	1 lb. 4 oz.
Celery, Raw, Diced	1/4 oz.	4 oz.	8 oz.	1 lb.
Cheese, Diced	—	5 1/2 oz.	11 oz.	1 lb. 6 oz.
Cheese, Grated	1/4 oz.	4 oz.	8 oz.	1 lb.
Cheese, Shredded	1/4 oz.	4 oz.	8 oz.	1 lb.
Chocolate, Grated	1/4 oz.	4 oz.	8 oz.	1 lb.
Chocolate, Melted	1/2 oz.	8 oz.	1 lb.	2 lb.
Cinnamon, Ground	1/4 oz.	3 1/2 oz.	7 oz.	14 oz.
Cloves, Ground	1/4 oz.	4 oz.	8 oz.	1 lb.
Cloves, Whole	3/16 oz.	3 oz.	6 oz.	12 oz.
Cocoa	3/16 oz.	3 1/2 oz.	7 oz.	14 oz.
Coconut, Macaroon, Packed	3/16 oz.	3 oz.	6 oz.	12 oz.
Coconut, Shredded, Packed	3/16 oz.	3 1/2 oz.	7 oz.	14 oz.
Coffee, Ground	3/16 oz.	3 oz.	6 oz.	12 oz.
Cornmeal	5/16 oz.	4 3/4 oz.	9 1/2 oz.	1 lb. 3 oz.
Cornstarch	1/3 oz.	5 1/3 oz.	10 1/2 oz.	1 lb. 5 oz.
Corn Syrup	3/4 oz.	12 oz.	1 lb. 8 oz.	3 lb.

Fig. 4-11 Approximate weights and measures of common foods.

Food Product	Tbsp.	Cup	Pt.	Qt.
Cracker Crumbs	1/4 oz.	4 oz.	8 oz.	1 lb.
Cranberries, Raw	—	4 oz.	8 oz.	1 lb.
Currants, Dried	1/3 oz.	5 1/3 oz.	11 oz.	1 lb. 6 oz.
Curry Powder	3/16 oz.	3 1/2 oz.	—	—
Dates, Pitted	5/16 oz.	5 1/2 oz.	11 oz.	1 lb. 6 oz.
Eggs, Whole	1/2 oz.	8 oz.	1 lb.	2 lb.
Egg Whites	1/2 oz.	8 oz.	1 lb.	2 lb.
Egg Yolks	1/2 oz.	8 oz.	1 lb.	2 lb.
Extracts	1/2 oz.	8 oz.	1 lb.	2 lb.
Flour, Bread	5/16 oz.	5 oz.	10 oz.	1 lb. 4 oz.
Flour, Cake	1/4 oz.	4 3/4 oz.	9 1/2 oz.	1 lb. 3 oz.
Flour, Pastry	5/16 oz.	5 oz.	10 oz.	1 lb. 4 oz.
Gelatin, Flavored	3/8 oz.	6 1/2 oz.	13 oz.	1 lb. 10 oz.
Gelatin, Plain	5/16 oz.	5 oz.	10 oz.	1 lb. 4 oz.
Ginger	3/16 oz.	3 1/4 oz.	6 1/2 oz.	13 oz.
Glucose	3/4 oz.	12 oz.	1 lb. 8 oz.	3 lb.
Green Peppers, Diced	1/4 oz.	4 oz.	1 lb. 8 oz.	1 lb.
Ham, Cooked, Diced	5/16 oz.	5 1/4 oz.	10 1/2 oz.	1 lb. 5 oz.
Horseradish, Prepared	1/2 oz.	8 oz.	1 lb.	2 lb.
Jam	5/8 oz.	10 oz.	1 lb. 4 oz.	2 lb. 8 oz.
Lemon Juice	1/2 oz.	8 oz.	1 lb.	2 lb.
Lemon Rind	1/4 oz.	4 oz.	8 oz.	1 lb.
Mace	1/4 oz.	3 1/4 oz.	6 1/2 oz.	13 oz.
Mayonnaise	1/2 oz.	8 oz.	1 lb.	2 lb.
Milk, Liquid	1/2 oz.	8 oz.	1 lb.	2 lb.
Milk, Powdered	5/16 oz.	1 1/4 oz.	9 oz.	1 lb. 2 oz.
Molasses	3/4 oz.	12 oz.	1 lb. 9 oz.	3 lb.
Mustard, Ground	1/4 oz.	3 1/4 oz.	6 1/2 oz.	13 oz.
Mustard, Prepared	1/4 oz.	4 oz.	8 oz.	1 lb.
Nutmeats	1/4 oz.	4 oz.	8 oz.	1 lb.
Nutmeg, Ground	1/4 oz.	4 1/4 oz.	8 1/2 oz.	1 lb. 1 oz.
Oats, Rolled	3/16 oz.	3 oz.	6 oz.	12 oz.
Oil, Salad	1/2 oz.	8 oz.	1 lb.	2 lb.
Onions	1/3 oz.	5 1/2 oz.	11 oz.	1 lb. 6 oz.
Peaches, Canned	1/2 oz.	8 oz.	1 lb.	2 lb.
Peas, Dry, Split	7/16 oz.	7 oz.	14 oz.	1 lb. 12 oz.
Pickle Relish	5/16 oz.	5 1/4 oz.	10 1/2 oz.	1 lb. 5 oz.
Pickles, Chopped	1/4 oz.	5 1/4 oz.	10 1/2 oz.	1 lb. 5 oz.
Pimientos, Chopped	1/2 oz.	7 oz.	14 oz.	1 lb. 12 oz.

Fig. 4-11 Approximate weights and measures of common foods (con't).

Food Products	Tbsp.	Cup	Pt.	Qt.
Pineapple, Diced	1/2 oz.	8 oz.	1 lb.	2 lb.
Potatoes, Cooked, Diced	—	6 1/2 oz.	13 oz.	1 lb. 10 oz.
Prunes, Dry	—	5 1/2 oz.	11 oz.	1 lb. 6 oz.
Raisins, Seedless	1/3 oz.	5 1/3 oz.	10 3/4 oz.	1 lb. 5 oz.
Rice, Raw	1/2 oz.	8 oz.	1 lb.	2 lb.
Sage, Ground	1/8 oz.	2 1/4 oz.	—	—
Salmon, Flaked	1/2 oz.	8 oz.	1 lb.	2 lb.
Salt	1/2 oz.	8 oz.	1 lb.	2 lb.
Savory	1/8 oz.	2 oz.	—	—
Shortening	1/2 oz.	8 oz.	1 lb.	2 lb.
Soda	7/16 oz.	7 oz.	—	—
Sugar, Brown, Packed	1/2 oz.	8 oz.	1 lb.	2 lb.
Sugar, Granulated	7/16 oz.	7 1/2 oz.	15 oz.	1 lb. 14 oz.
Sugar, Powdered	5/16 oz.	4 3/4 oz.	9 1/2 oz.	1 lb. 3 oz.
Tapioca, Pearl	1/4 oz.	4 oz.	8 oz.	1 lb.
Tea	1/6 oz.	2 1/2 oz.	5 oz.	10 oz.
Tomatoes	1/2 oz.	8 oz.	1 lb.	2 lb.
Tuna Fish, Flaked	1/2 oz.	8 oz.	1 lb.	2 lb.
Vanilla, Imitation	1/2 oz.	8 oz.	1 lb.	2 lb.
Vinegar	1/2 oz.	8 oz.	1 lb.	2 lb.
Water	1/2 oz.	8 oz.	1 lb.	2 lb.

Fig. 4-11 Approximate weights and measures of common foods (con't).

—————— ACHIEVEMENT REVIEW ——————
—— WEIGHING AND PRICING FOOD PORTIONS USING THE PORTION SCALE ——

1. Determine the cost of a portion of cooked roast beef if the dial pointer on the above portion scale points to the second mark beyond the 4 and the cost of 1 pound of cooked roast beef is $4.50.

2. Determine the cost of a chop steak if the dial pointer on the above portion scale points to the second mark beyond the 5 and the cost of 1 pound of lean ground beef is $1.45.

3. Determine the cost of a portion of cooked turkey if the dial pointer on the above portion scale points to the third mark beyond the 3 and the cost of 1 pound of cooked turkey is $3.65.

4. Determine the cost of a portion of corn beef if the dial pointer on the above portion scale points to the third mark beyond the 2 and the cost of 1 pound of cooked corn beef is $3.75.

5. Determine the cost of a sirloin steak if the dial pointer on the above portion scale points to the second mark beyond the 12 and the cost of 1 pound of trimmed sirloin is $6.25.

6. Determine the cost of a portion of cooked ham if the dial pointer on the above portion scale points to the first mark beyond the 3 and the cost of 1 pound of cooked ham is $1.85.

7. Determine the cost of a veal cutlet if the dial pointer on the above portion

scale points to the third mark beyond the 5 and the cost of 1 pound of trimmed leg of veal is $5.25.

8. Determine the cost of a portion of roast loin of pork if the dial pointer on the above portion scale points to the second mark beyond the 4 and the cost of 1 pound of cooked pork loin is $2.95.

9. Determine the cost of a portion of roast sirloin of beef if the dial pointer on the above portion scale points to the first mark beyond the 6 and the cost of 1 pound of roast sirloin is $4.65.

10. Determine the cost of a filet mignon if the dial pointer on the above portion scale points to the second mark beyond the 8 and the cost of 1 pound of trimmed beef tenderloin is $6.95.

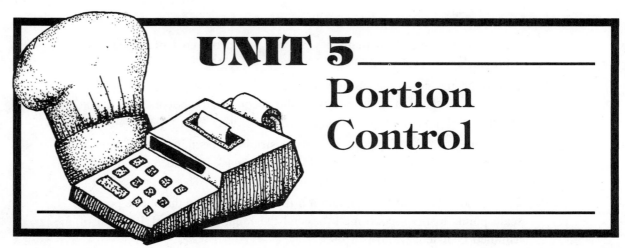

UNIT 5
Portion Control

Controlling the size of portions is essential in all food service establishments so that a profit can be made. The best way to control portions is to use standardized recipes which state the number of servings a preparation will produce. However, a standardized recipe gives only the stated number of portions if the servings are uniform in size. To ensure uniform servings, the production crew must be instructed in the use of scoops, ladles, spoons, scales, and similar devices when dishing up.

Another key to a successful portion control program is intelligent buying. Buy foods in sizes that portion well. Work out buying specifications that suit the portion need. For example, cooked smoked ham can be purchased in many types and sizes. Purchase the kind that will produce a ham steak the diameter desired and one that produces little or no waste. Most link sausage such as hamilton mett, pork links, and frankfurters can be purchased at a certain number (6, 8, or 10) to each pound. Purchase the count per pounds that best suits the portion requirement. Select veal, pork, lamb, and beef ribs and loins that provide the size chop or slice desired. Some fabricated foods can be purchased ready to cook,

and are purchased for absolute portion control. This is another controlling device to consider. Fish fillets, steaks, lamb, veal, pork chops, and beef steaks are all cut to the exact ounce desired, figure 5-1. The cost per pound is much higher, but many food service operators feel that the

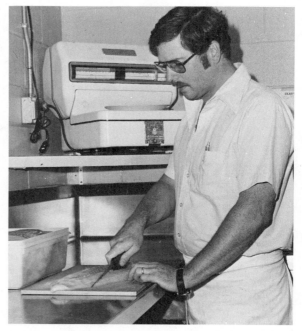

Fig. 5-1 Fish fillets are cut to the exact serving portions desired.

final cost is lower considering the following factors:

- No leftovers
- Less storage required
- No waste
- No cutting equipment to purchase
- Less labor cost

When portioning food for a particular establishment, remember that portions can be too large as well as too small. Therefore, before a policy is established, the manager, as well as the chef, should know the customers. This knowledge can be acquired by carefully observing the plates that are brought into the dishwashing area. Too much uneaten food left in a bowl or on a plate indicates that a portion is too large, or that the quality of the food does not satisfy the customer. In either case, the situation must be corrected to improve customer satisfaction and to control food cost. Too small a portion is usually indicated by plates and bowls that are scraped entirely clean. A satisfied guest usually leaves a very small amount of food on the plate or in the bowl.

Controlling portions in a food service establishment makes menu pricing, purchasing, and food production more accurate. It also reduces the food cost. When portioning food in ounces per serving, figure 5-2, it is easy to find how much raw food is needed and how much should be prepared for a specific number of people. By finding the cost per ounce, a total cost is easy to calculate. Many chefs and managers become tired of hearing employees ask "How much should I prepare?" By observing portion control charts posted in the preparation area, and by doing

Fig. 5-2 Cook portioning chopped steaks on a portion scale.

some simple figuring, employees could answer their own questions.

Once a portion policy is established, it should be posted in the kitchen. A typical portion chart is shown in figure 5-3.

A point to remember when figuring portion sizes is that the average human stomach can only hold approximately 2½ pounds of solid and liquid food comfortably. Therefore, oversized portions do not make customers satisfied, and usually create more waste. The intelligent restaurant operator figures portion sizes so that the customer has room left for dessert, figure 5-4. An example of how portion sizes should add up is given in figure 5-5.

Steaks	
Chateaubriand	16 oz.
Filet Mignon	6 oz.
Minute	6 oz.
Sirloin	10 oz.
NY Strip Sirloin	16 oz.
T-bone	12 oz.
Club	10 oz.
Porterhouse	14 oz.
Baby T-bone	6 oz.
Beef Tenderloin	8 oz.
Salisbury	8 oz.
Ham	6 oz.
Veal Steak	6 oz.
Lamb Steak	7 oz.

Chops and Cutlets	
Pork Chops	2-3½ oz. each
Lamb Chops	2-3½ oz. each
Veal Chops	1-6 oz.
English Lamb Chop	6 oz.
Veal Cutlet	6 oz.
Pork Cutlet	6 oz.
Escallop of Veal	7 oz.
Noisette of Lamb	2-3½ oz. each
Pork Tenderloin	7 oz.
Beef Tournedos	2-4 oz. each

Poultry	
Fried Chicken	½ fryer — 3 ½ lb. chicken
Broiled Chicken	½ broiler — 2 lb. chicken
Roast Chicken	½ chicken 3 lb. chicken
Roast Turkey	2 ½ oz. white meat — 3 oz. dark meat
Turkey Steak	5 oz. white meat
Boneless Turkey Wings	2 wings
Chicken a la King	6 oz.
Chicken Pot Pie	6 oz. plus crust
Chicken a la Maryland	½ fryer — 1 oz. bacon — 2 oz. cream sauce — 2 oz. cornfritters — 2 croquettes 6 oz.
Chicken Cutlet	2 cutlets — 6 oz.
Roast Duck	8 oz.
Roast Squab	1 bird
Roast Boneless Breast of Chicken	5 oz. breast
Baked Stuffed Chicken Leg	1 leg 3 oz. stuffing

Seafood	
Lobster, Broiled Whole	24 oz.
Lobster Newburg	5 oz. meat
Lobster Crab	5 oz. meat
Fried Shrimp	6 jumbo — 8 medium
Shrimp Newburg	7 medium
Sauteed Shrimp	6 jumbo — 8 medium
Softshell Crabs	2 crabs
Clam Roast	8 cherrystone
Steamed Clams	8 cherrystone
Fried Clams	8 cherrystone
Fried Oysters	7 select
Oyster Stews	6 select
Fried Scallops	8 small — 6 large
Sauteed Scallops	8 small — 6 large
Halibut	7 oz.
Cod	7 oz.
Spanish Mackerel	7 oz.
Pampano	7 oz.
Red Snapper	6 oz.
Frog Legs	8 oz.
White Fish	7 oz.
Lake Trout	7 oz.
Rainbow Trout	8 oz.
Brook Trout	8 oz.
Smelts	6 fish, about 7 oz.
Salmon	7 oz.
Shad Roe	4 oz.
English and Dover Sole	7 oz.

Roasted Meats	
Roast Rib of Beef	8 oz.
Roast Tenderloin of Beef	6 oz.
Roast Sirloin of Beef	6 oz.
Roast Round of Beef	5 oz.
Roast Leg of Lamb	5 oz.
Roast Loin of Pork	6 oz.
Roast Leg of Veal	5 oz.
Roast Fresh Ham	6 oz.
Baked Ham	6 oz.

Fig. 5-3 Standardized portion chart.

Stews, Blanquettes, Hashes, Etc.	
Beef Goulash	7 oz.
Beef Stew	7 oz.
Veal Blanquette	6 oz.
Lamb Blanquette	7 oz.
Lamb Stew	7 oz.
Veal Stew	7 oz.
Oxtail Stew	10 oz.
Roast Beef Hash	6 oz.
Corned Beef Hash	6 oz.
Chicken Hash	6 oz.
Beef Stroganoff	7 oz.
Beef a la Deutsch	7 oz.

Potato Preparations	
Baked	6 oz.
Au Gratin	4 oz.
Delmonico	4 oz.
French Fried	5 oz.
Mashed	5 oz.
Julienne	4 oz.
Lyonnaise	5 oz.
Croquette	5 oz.
Hash Brown	5 oz.
Escallop	4 oz.
Candied Sweet	5 oz.

Vegetables	
Asparagus, Spears	4 or 5 spears
Asparagus, Cut	4 oz.
Beans, Limas	4 oz.
Beans, String	4 oz.
Beans, Wax	4 oz.
Beets	4 oz.
Brussels Sprouts	5 oz.
Cabbage	5 oz.
Cauliflower	5 oz.
Carrots	4 oz.
Corn on Cob	1 cob
Corn, Whole Kernel	4 oz.
Corn, Cream Style	5 oz.
Mushrooms, Whole	4 oz.
Onions	5 oz.
Peas	4 oz.
Rice	4 oz.
Squash	4 oz.
Succotash	3 oz.
Tomatoes, Stewed	4 oz.
Eggplant	4 oz.

Desserts	
Baked Alaska	1 slice — 8 per Alaska
Coupes	5 oz.
Cake	1 slice — 8 per cake
Ice Cream	4 oz.
Jubilee	5 oz. ice cream 2 oz. cherries
Parfaits	5 oz. ice cream 3 oz. sauce
Pie	1 slice — 6 per pie
Pudding	5 oz.
Sherbets	4 oz.

Salads	
Cole Slaw	4 oz.
Garden Salad	5 oz.
Ham Salad	5 oz.
Julienne	5 oz.
Macaroni	4 oz.
Potato	5 oz.
Toss	5 oz.
Waldorf	5 oz.

Fig. 5-3 Standardized portion chart (con't)

COST PER SERVING

To find the cost per serving, the total weight of the item is converted into ounces, and divided into the total cost to find the cost of one ounce. The cost of one ounce is multiplied by the number of ounces being served.

Example: A 5-pound box of frozen lima beans costs $3.50. How much does a 4-ounce serving cost?

$$\begin{array}{r} \$\ .043 \quad \text{Cost of 1 ounce} \\ 80\overline{)\$3.500} \\ \underline{3\ 20} \\ 300 \\ \underline{240} \end{array}$$

Fig. 5-4 Proper portion size is simplified when sheet cakes are decorated according to individual portions.

The division is carried to three places to the right of the decimal point. Remember, the third digit is the mill, or ¹⁄₁₀ of a cent.

The cost per ounce ($.043) is now multiplied by the number of ounces in each serving. In this case, 4 ounces are contained in one serving.

```
  $.043    Cost per ounce
×     4    Ounces per serving
  $.172    Cost for each 4-ounce
           serving
```

Of course, if a cost per pound is given rather than a total, the number of pounds given

Appetizers	4 oz.
Salad	4 oz.
Entree	8 oz.
Potato	4 oz.
Vegetable	4 oz.
Bread and Butter	3 oz.
Dessert	6 oz.
Beverage	7 oz.
	40 oz. = 2 ½ lb.

Fig. 5-5 An example of how portion sizes should add up.

must be multiplied by the cost per pound to find a total cost.

Example: Find the cost of a 3-ounce serving of succotash (mixture of two vegetables) if the following ingredients are used:

5-pound box lima beans @ $0.34 per pound
2½-pound box corn @ $0.31 per pound

```
  $ .34    Cost per pound
×     5    Pounds contained in box
  $1.70    Cost for the 5-pound box
```

```
   $.31    Cost per pound
×   2.5    Pounds contained in box (2½
    155      converted to decimal form)
     62
  $.775    Cost for the 2½-pound box
```

```
  $1.70    Cost of lima beans
+  .775    Cost of corn
  $2.475   Total cost of succotash
```

To complete the problems, follow the same steps explained in the previous example.

7½ pounds combined weight of both vegetables

16 ounces in one pound

multiply 16 × 7½ = 120 ounces

```
      $.020   Cost per ounce
120)2.475
    2 40
      75
```

```
   $.02    Cost per ounce
×     3
   $.06    Cost for each 3-ounce serving
```

ACHIEVEMENT REVIEW
COST PER SERVING

Round answers to the nearest cent.

1. A 2½-pound box of frozen corn costs $.96. How much does a 4-ounce serving cost?

2. A 5-pound box of frozen peas costs $1.60. How much does a 3½-ounce serving cost?

3. When preparing succotash, a 2½-pound box of frozen corn and a 5-pound box of frozen lima beans are used. Both cost $.34 per pound. How much does a 3-ounce serving cost?

4. A 2½-pound box of frozen peas and onions costs $.36 per pound. How much does a 3½-ounce serving cost?

5. Find the cost of a 4-ounce serving of Kentucky succotash if the following ingredients are used.

   2½-pound box frozen green beans @ $.36 per pound
   2½-pound box frozen corn   @ $.34 per pound

6. If frozen asparagus costs $3.29 for a 5-pound box, how much does a 3-ounce serving cost?

7. Find the cost of a 3½-ounce serving of mixed vegetables if the following items are used.

2½-pound box of frozen peas	@ $.36 per pound
5-pound box of frozen lima beans	@ $.38 per pound
2½-pound box of frozen carrots	@ $.32 per pound
5-pound box of frozen corn	@ $.30 per pound

8. Find the cost of a 4-ounce serving of mixed vegetables if the following items are used.

5-pound box frozen peas	@ $.36 per pound
2½-pound box frozen corn	@ $.30 per pound
5-pound box frozen string beans	@ $.35 per pound
2½-pound box frozen carrots	@ $.32 per pound
2½-pound box frozen lima beans	@ $.38 per pound

9. Find the cost of a 3½-ounce serving of peas and carrots if the following items are used.

    5-pound box frozen peas costing $1.95
    2½-pound box frozen carrots costing $.85

10. If a 2½-pound box of frozen cut broccoli costs $.94, what is the cost of a 3½-ounce serving?

PORTIONING WITH SCOOPS OR DIPPERS

Scoops or dippers, as they are sometimes called, are used to serve and portion certain foods. A scoop or dipper has a metal bowl or cup of known capacity, with an extended handle. A movable strip of metal on the inside of the bowl is attached to a thumb-operated lever that, when operated, will release the item it holds. See figure 5-6. Scoops are numbered on the movable metal strip to indicate the size of the metal cup—the larger the number, the smaller the cup—the number indicates the number of scoopfuls it will take to make one quart. Figure 5-7 relates each scoop number to its approximate capacity in ounces. Figure 5-8 relates the scoop number to the approximate content of each scoop size in cups or tablespoons. Scoops or dippers can be used to portion meat patties, baked rice, bread dressing, fritter batter, muffin batter, potato preparations, croquette mixtures, and salads.

To find the number of servings a particular amount of food will produce when portioning with a scoop or dipper, divide the amount con-

Scoop or Dipper Number	Approximate Weight
8	5 ounces
10	4 ounces
12	3 ounces
16	2–2$\frac{1}{2}$ ounces
20	1$\frac{2}{3}$ ounces
24	1$\frac{1}{2}$ ounces
30	1$\frac{1}{4}$ ounces
40	1 ounce

Fig. 5-7 Scoop or dipper sizes and approximate weights.

tained in the scoop or dipper into the amount being portioned.

How many servings can be obtained from 14 pounds of bread dressing if a No. 12 scoop is used to portion?

Convert the 14 pounds to 224 ounces (14 × 16 = 224 ounces). A No. 12 scoop contains 3 ounces.

$$
\begin{array}{r}
74 \text{ servings} \\
3 \text{ oz.} \overline{)224.} \text{ oz.} \\
21 \\
\hline
14 \\
12 \\
\hline
2
\end{array}
$$

Scoop or Dipper Number	Approximate Measures
8	$\frac{1}{2}$ cup
10	$\frac{2}{5}$ cup
12	$\frac{1}{3}$ cup
16	$\frac{1}{4}$ cup
20	3$\frac{1}{5}$ tablespoons
24	2$\frac{2}{3}$ tablespoons
30	2$\frac{1}{5}$ tablespoons
40	1$\frac{3}{5}$ tablespoons

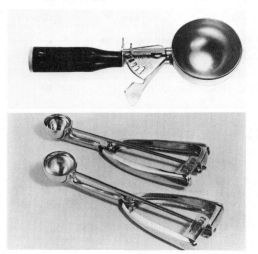

Fig. 5-6 Two examples of food scoops. (Courtesy of Hamilton Beach)

Fig. 5-8 Scoop or dipper sizes and approximate measures.

There is no need to carry the problem further, because figures on the right of the decimal point represent a part of a full scoop.

Example:
How many servings can be obtained from 2 gallons of ice cream if a No. 8 scoop is used?

Convert 2 gallons to 32 cups (1 gallon contains 16 cups; hence 16 × 2 = 32 cups). A No. 8 scoop contains ½ cup (see figure 5-8): Thus

$$32 \div \frac{1}{2} = \frac{32}{1} \times \frac{2}{1} = 64 \text{ servings}$$

It was necessary to convert the bread dressing to ounces and the ice cream to cups because one can only divide like things.

ACHIEVEMENT REVIEW
PORTIONING WITH SCOOPS OR DIPPERS

1. How many servings can be obtained from 3 gallons of mashed potatoes if a No. 12 scoop is used?

2. How many corn fritters can be obtained from 3 pounds of batter if a No. 24 scoop is used?

3. Determine how many servings can be obtained from 9 pounds of bread pudding if a No. 10 scoop is used?

4. How many servings can be obtained from 5 quarts of strawberry mousse if a No. 12 scoop is used?

5. How many individual salads can be obtained from 6 pounds of tuna salad if a No. 10 scoop is used to portion?

6. How many clam fritters can be obtained from 6 pounds of batter if a No. 24 scoop is used to portion?

7. How many corn muffins can be obtained from 2 gallons of corn muffin batter if a No. 12 scoop is used?

8. How many servings can be obtained from ½ gallon of cold pack cheese if a No. 16 scoop is used to portion?

9. How many hush puppies can be obtained from 2 pounds of batter if a No. 30 scoop is used to portion?

10. How many servings can be obtained from 12 pounds of mashed potatoes if a No. 8 scoop is used to portion?

FIGURING AMOUNTS TO PREPARE

In the commercial kitchen, it is constantly necessary to figure how many cans, boxes, or packages of certain food items are needed or must be opened in order to have enough to serve a given number of people. This problem is solved by multiplying the number of people to be served by the portion size. This gives the number of ounces needed. Next, the weight of the can, box, or package is converted into ounces and the given amount is divided into the number of ounces needed. If a remainder results from the division, an additional container must be opened. By utilizing mathematics in this situation, guesswork can be eliminated and food preparation can be controlled.

Example: How many No. 10 cans of corn are needed to serve a party of 161 people if each person is to receive a 3-ounce serving and each can contains 4 pounds 5 ounces?

$$\begin{array}{r} 161 \\ \times\ \ \ 3 \\ \hline 483 \end{array}$$

161 people to be served
× 3 ounces per serving
483 ounces needed to serve 161 people

Convert 4 pounds 5 ounces to 69 ounces (content of one can)

$$\begin{array}{r} 7 \\ 69\overline{)483} \\ 483 \\ \hline \end{array}$$ No. 10 cans needed

The above example shows that 7 No. 10 cans are needed to serve each of the 161 people a 3-ounce serving. If, when dividing the content of the can into the number of ounces needed, a remainder results, an additional can (making the total 8) is required.

ACHIEVEMENT REVIEW
AMOUNTS TO PREPARE

1. A 4-ounce serving of peas is to be served to each of 170 people. How many boxes of frozen peas should be cooked if each box weighs 2½ pounds?

2. A 3½-ounce serving of applesauce is served to each of 95 people. How many No. 2½ cans of applesauce are needed if each can weighs 2 pounds 6 ounces?

3. A 3-ounce serving of wax beans is served to each of 50 people. How many No. 303 cans of wax beans are needed if each can weighs 1 pound 2 ounces?

4. How many No. 2½ cans of pork and beans are required to serve 75 people if each can weighs 1 pound 3 ounces and each serving is 5 ounces?

5. A 3-ounce serving of peas is to be served to each of 100 people. How many boxes of frozen peas are needed if each box weighs 5 pounds?

6. How many No. 10 cans of green beans are needed to serve a party of 250 people if each person is to receive a 3½-ounce serving and each can weighs 4 pounds 6 ounces?

7. How many boxes of frozen chopped spinach are needed to serve a party of 185 people, if each person is to receive a 3-ounce serving and each box weighs 4 pounds 6 ounces.

8. How many No. 2½ cans of diced beets are needed to serve a party of 45 people if each person is to receive a 4½-ounce serving and each can weighs 2 pounds 7 ounces?

9. How many 5-pound boxes of frozen lima beans are needed to serve a party of 350 people, if each person is to receive a 4-ounce serving?

10. A 4½-ounce serving of sliced carrots is served to each of 140 people. How many cans are needed if each can weighs 5 pounds 10 ounces?

APPROXIMATE NUMBER OF SERVING PORTIONS

Finding approximately how many servings can be acquired from a given amount of food (figure 5-9) or how much of a certain meat, fish, vegetable, or liquid food product should be ordered are other portion control problems. For example, how many 10-ounce strip steaks can be cut from a sirloin? How many 2-ounce meatballs can be acquired from a certain amount of ground beef? How many pounds of fish steaks must be ordered for a party of 90 people? How many gallons of orange juice must be ordered to serve a party of 200 people?

The arithmetic involved in these problems

Fig. 5-9 Student is portioning meat to determine how many servings it will produce.

is quite simple, but is an essential part of a food service operation. This arithmetic must be accurate to keep inventories at a minimum, to control waste, and to maintain an effective portion control program.

To find out how many servings can be obtained from a given amount, the actual usable amount must first be established (that is, waste must be eliminated). Waste such as bone, fat, and skin cannot be converted into serving portions so it must be subtracted from the original weight of the product. The size of the serving portion is then divided into the actual usable amount to give the approximate number of servings it will produce.

Example: A 17-pound strip sirloin of beef is purchased and 3 pounds 12 ounces are lost through boning and trimming. How many 10-ounce strip sirloin steaks can be cut from the sirloin?

Convert 17 pounds to ounces.

$$
\begin{array}{r}
17 \\
\times\ 16 \\
\hline
102 \\
17 \\
\hline
272 \quad \text{Ounces contained in 17 pounds.}
\end{array}
$$

3 pounds 12 ounces were lost through boning and trimming. Convert this to ounces.

$$
\begin{array}{r}
16 \\
\times\ \ 3 \\
\hline
48 \quad \text{ounces} \\
+\ \ 12 \quad \text{ounces} \\
\hline
60 \quad \text{Total ounces lost}
\end{array}
$$

$$
\begin{array}{r}
272 \quad \text{ounces—original amount} \\
-\ \ \ 60 \quad \text{ounces—lost} \\
\hline
212 \quad \text{actual usable amount}
\end{array}
$$

$$\begin{array}{r} 21 \\ 10 \text{ ounces } \overline{)212} \\ \underline{20} \\ 12 \\ \underline{10} \end{array}$$ Steaks
ounces

Each steak is to weigh 10 ounces, so the weight of each steak is divided into 212 ounces, the actual usable amount. This shows that 21 steaks can be cut from the strip sirloin that weighed 17 pounds.

ACHIEVEMENT REVIEW
FINDING APPROXIMATE NUMBER
OF SERVING PORTIONS

1. How many strip steaks can be cut from a sirloin of beef weighing 21 pounds, if 3 pounds 8 ounces are lost through trimming and boning, and each steak is to weigh 10 ounces?

2. A 7-pound beef tenderloin is trimmed and 12 ounces are lost. How many 8-ounce filet mignons can be cut from the tenderloin?

3. How many 5-ounce pork chops can be cut from a pork loin weighing 14 pounds if the tenderloin, which is removed, weighed 10 ounces, and 4 pounds 6 ounces are lost through boning and trimming?

4. How many orders of meatballs can be obtained from 40 pounds of ground beef, if each meatball is to weigh 2 ounces and two meatballs are served per order?

5. How many orders of Swedish meatballs can be obtained from 38 pounds of ground pork and veal, if each meatball weighs 1½ ounces and 5 meatballs are served with each order?

6. How many 6-ounce Swiss steaks can be cut from a beef round weighing 42 pounds if 5 pounds 4 ounces are lost in boning and trimming?

7. A 5-pound 8-ounce beef tenderloin is trimmed and 12 ounces are lost. How many beef tenderloin steaks can be cut from the tenderloin if each steak is to weigh 6 ounces?

8. When preparing pork sausage, 18 pounds of pork picnic is purchased. One third of that amount is lost through boning and trimming. How many 5-ounce patties can be obtained?

9. How many 5-ounce glasses of orange juice can be obtained from 1 gallon of orange juice? (1 qt. = 32 oz.)

10. A 70-pound halibut is purchased; 4 pounds 5 ounces are lost through boning and 3 pounds 6 ounces are lost through skinning and trimming. How many halibut steaks can be obtained if each steak is to weigh 6 ounces?

ORDERING FOOD

When ordering food for a specific number of people, the amount to order can be found by multiplying the amount of the serving portion by the number of people to be served. This gives the number of ounces needed. Next, convert the common purchasing quantity into ounces and divide this amount into the number of ounces needed (of course, in the case of meat, fish, etc., consideration must be given to the amount that may be lost through boning and trimming).

Example: How many pounds of ground beef should be ordered if 45 people are to be served and each person is to receive a 6-ounce portion?

$$
\begin{array}{r r l}
45 & & \text{Number of people being served} \\
\times\ 6 & & \text{Serving portion—ounces} \\
\hline
270 & & \text{Number of ounces to} \\
& & \text{serve 45 people}
\end{array}
$$

$$
16 \frac{14}{16} = 17 \text{ pounds must be ordered}
$$

$$
\begin{array}{r}
16\overline{)270} \\
\underline{16} \\
110 \\
\underline{96} \\
14
\end{array}
$$

A total of 270 ounces will be required to serve the 45 people. The common purchasing quantity for meat is pounds. Since there are 16 ounces in a pound, 16 is divided into the number of ounces needed. The result is 16, and $^{14}/_{16}$ remaining, so the actual number of pounds to be ordered must be 17. The remainder indicates that 16 pounds will not produce enough portions to serve 45 people.

ACHIEVEMENT REVIEW
FINDING AMOUNTS TO ORDER

1. Roast sirloin of beef is being served to a party of 250 people. Each person is to receive 9 ounces of meat. How many pounds of meat must be ordered?

2. Referring to problem No. 1, if each strip loin weighs 14 pounds, how many sirloins must be ordered?

3. Swiss steak is being served to a party of 120 people. The steaks are cut into 6-ounce portions. If 3 pounds 12 ounces are allowed for trimming and boning, how many pounds of round must be ordered?

4. Salisbury steak is to be served to a party of 185 people. Each portion is to weigh 6 ounces. How many pounds of ground beef must be ordered?

5. How many pounds of pork sausage should be ordered for 65 people, if each person is to receive two 3-ounce patties?

6. How many gallons of orange juice must be ordered for a party of 180 people, if each person is to receive a 5-ounce glass? (1 qt. = 32 oz.)

7. How many pounds of bacon should be ordered when serving a party of 85 people, if each person is to receive 3 slices of bacon and there are 32 slices in each pound?

8. When preparing a breakfast for 230 people, how many pounds of sausage must be ordered, if each person is to receive 3 sausages and there are 10 sausages in each pound?

9. How many pounds of short ribs should be ordered when preparing for 90 people, if each person is to receive a 12-ounce portion?

10. How many pounds of ground beef must be ordered to serve spaghetti and meatballs to 140 people if each person is to receive two 2½ ounce meatballs?

PURCHASING FRESH FISH

The quantity of fresh fish to purchase depends on three things: the number of people being served, the size of portion, and the market form. The following is a suggested guide associating the market form to the amount to purchase:

Type of Fish	Amount per Person
Fish Sticks, Steaks, and Fillets	⅓ pound
Dressed Fish	½ pound
Drawn Fish	¾ pound
Whole Fish or Fish in the Round (just as it comes from the water)	1 pound

Example 1: How many pounds of fish steaks must be ordered for serving a party of 75 people?

According to the guide, ⅓ pound per person is needed. Hence,

$$\frac{1}{3} \times \frac{75}{1} = 25 \text{ pounds}$$

Example 2: How many pounds of dressed fish should one order when preparing for a group of 76 people?

Again using the guide, we have

$$\frac{1}{2} \times \frac{76}{1} = 38 \text{ pounds}$$

Example 3: How many pounds of drawn fish should one order when preparing for a party of 48 people?

$$\frac{3}{4} \times \frac{48}{1} = 36 \text{ pounds}$$

ACHIEVEMENT REVIEW
PURCHASING FRESH FISH

1. How many pounds of drawn fish should be ordered when preparing for a party of 96 people?

2. How many pounds of fish steaks should be ordered when preparing for a group of 66 people?

3. How many pounds of fish fillets should be ordered when preparing for a party of 96 people?

4. How many pounds of fish sticks should be purchased when preparing for a group of 84 people?

5. How many pounds of drawn fish should be purchased when preparing for a party of 122 people?

6. How many pounds of dressed fish should be ordered when preparing for a party of 120 people?

7. How many pounds of fish steaks should be ordered when preparing for a group of 126 people?

8. How many pounds of fish fillets should be ordered when preparing for a party of 114 people?

9. How many pounds of fish sticks should be purchased when preparing for a group of 72 people?

10. How many pounds of drawn fish should be ordered when preparing for a party of 52 people?

COSTING MEAT AND FISH PORTIONS

Meat and fish are usually the most popular entrees on a menu, but also the most expensive. The average commercial establishment spends approximately 25 to 35 percent of its food dollar for meat and fish. Therefore, it is very important that the food service operator be constantly aware of portion cost, figure 5-10.

To find the cost of a portion, the total cost of the amount purchased must first be established. This is done by multiplying the price per pound by the number of pounds purchased. The amount purchased is then converted into ounces by multiplying the number of pounds purchased by 16 (16 ounces to one pound). The amount lost through boning, trimming, or shrinkage is converted into ounces and subtracted from the original amount purchased. This gives the actual usable amount. The actual usable amount is divided into the total

Fig. 5-10 This student is portioning Swiss steaks to control serving size and portion cost.

cost. (Carry the division three places to the right of the decimal.) This gives the cost of one ounce of cooked meat. The cost of one ounce is multiplied by the number of ounces contained in one portion.

Example: An 8-pound leg of lamb costs $1.28 per pound. A total of 12 ounces is lost in boning and 2 pounds through shrinkage during the roasting period. How much does a 4-ounce serving cost?

$1.28	Price per pound
× 8	Number of pounds purchased
$10.24	Total cost

16	Ounces in one pound
× 8	Number of pounds purchased
128	Number of ounces purchased

128	Number of ounces purchased
− 44	Number of ounces lost (boning & shrinkage)
84	Actual usable amount

```
       $  .121    Cost of one ounce of
   84)$10.240        cooked meat
       8 4
       1 84
       1 68
         160
          84
```

$ .121	Cost of one ounce
× 6	Ounces per serving
$.726	= Cost of each 6-ounce serving

Once the actual cost of a serving is established, it is easy to detrmine a selling price.

ACHIEVEMENT REVIEW
COSTING MEAT AND FISH PORTIONS

1. If a 22-pound leg of veal costs $1.85 per pound, and 5 pounds 4 ounces are lost through trimming and boning, how much does a 5-ounce veal cutlet cost?

2. A 24-pound rib of beef costs $1.49 per pound; 3 pounds are lost through trimming and ⅓ of the remaining weight is lost through shrinkage when it is roasted. How much does a 6-ounce serving cost?

3. A 50-pound round roast costs $62.50; 5 pounds are lost through trimming and 5 pounds through shrinkage when it is roasted. How much does one pound of cooked meat cost?

4. A 6-pound 8-ounce beef tenderloin costs $14.40; 1 pound 3 ounces are lost through trimming. How much does an 8-ounce filet mignon cost?

5. If a 20-pound rib of beef costing $28.00 is purchased oven ready and one-fourth of it is lost during roasting, how much does a 5-ounce serving cost?

6. A 9-pound leg of lamb costing $1.59 per pound is roasted. When the roast is removed from the oven, only two-thirds of the original amount is left. How much does a 4½-ounce serving cost?

7. Roast sirloin of beef is served to a party. Each sirloin weighs 12 pounds and costs $1.35 per pound. If 17 portions are cut from the sirloin, what is the cost per serving?

8. A 10-pound box of halibut steaks is ordered costing $1.35 per pound. If each steak weighs 5 ounces, what is the cost of each steak?

9. A 14-pound pork loin costs $1.29 per pound. A 10-ounce tenderloin is removed from the loin and 2 pounds 12 ounces are lost through boning and trimming. How much does a 5-ounce pork chop cost?

10. A 9-pound saddle of lamb costs $1.69 per pound; 1 pound 6 ounces are lost in trimming. How much does a 4-ounce lamb chop cost?

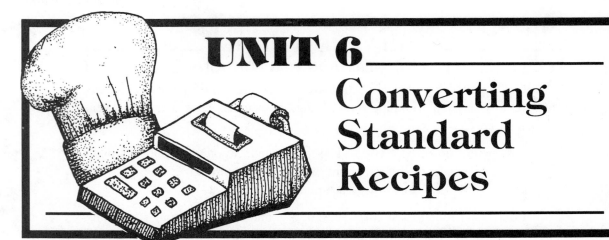

UNIT 6
Converting Standard Recipes

A standard recipe (or formula) produces a standard yield. Often, this standard yield is too much or too little, which means that the recipe or formula must be converted to yield the amount needed, figure 6-1.

To obtain the desired amounts in a standard recipe, a working factor must be found. The quantity of each ingredient is then multiplied by the working factor.

If the yield desired is greater than the amount yielded by the standard recipe, divide the standard yield of the recipe into the amount desired. The result is the working factor.

Example:
$$\begin{array}{l} \text{Recipe Yield: } 75 \\[4pt] \underline{3}\quad\text{Working factor} \\ 75\overline{)225}\quad\text{Amount desired} \\ \underline{225} \end{array}$$

This problem shows that the amount desired is 3 times the amount yielded by the standard recipe. Therefore, multiply the quantity of each ingredient in the recipe by 3 to obtain the desired amount. The example shows that the working factor is a whole number which makes multiplication simple, however, this is not always the case. Sometimes the working factor is a mixed decimal fraction.

Example:
$$\begin{array}{l} \text{Recipe Yield: } 75 \\[4pt] \underline{4.4}\quad\text{Working factor} \\ 75\overline{)330.0}\quad\text{Amount desired} \\ \underline{300} \\ \ \ 30\ 0 \\ \ \ \underline{30\ 0} \end{array}$$

This example shows that the amount desired is 4.4 times the amount yielded by the standard recipe. Therefore, multiply the quantity of each

Fig. 6-1 Student converting a recipe to determine the amounts needed to produce the required number of servings.

ingredient in the recipe or formula by 4.4 (the working factor) to obtain the desired amount.

If the yield desired is smaller than the amount yielded by the standard recipe, only a fraction of the original yield is needed. Therefore, place the amount desired over the amount yielded by by the standard recipe to form a fraction. Simplify the fraction to lowest terms to find the working factor and multiply the quantity of each ingredient in the recipe by the fraction to get the desired amounts.

Example: A certain standard recipe yields 40 servings. Only 30 servings are desired.

Place 30 over 40 $\frac{(30)}{(40)}$, simplify to ¾, and proceed to multiply the quantity of each ingredient in the recipe or formula by ¾, the working factor.

In commercial restaurants and bakeries, foods are usually produced in large quantities which must be very accurate. Therefore, the formulas are written in pounds and ounces. To make the arithmetic simple when converting recipes, change the pounds to ounces before starting to multiply. In this way, it is only necessary to

multiply ounces. After multiplying, convert the product back to pounds and ounces.

Example: To multiply 2 pounds 12 ounces by ¾, convert 2 pounds 12 ounces to 44 ounces. Then proceed to multiply.

$$\frac{3}{\overset{}{\underset{1}{4}}} \times \frac{\overset{11}{44}}{1} = 33 \text{ ounces}$$

Convert back to pounds and ounces:
$$33 \text{ ounces} = 2 \text{ pounds } 1 \text{ ounce}$$

When working a recipe or formula, some common sense must also be used. For example, the recipe may not account for how fresh or old the spices are or how hot the room is when mixing a yeast dough. Common sense must be applied when converting recipes or formulas because it may not be practical to increase or decrease the quantity of each ingredient by the same rate. Many spices cannot be increased or decreased at the same rate as other ingredients. This may also be true of salt, garlic, and sugar in certain situations. Use good judgment in these situations and carry out tests before deciding on amounts.

ACHIEVEMENT REVIEW
CONVERTING STANDARD RECIPES

1. The following recipe yields 50 portions of curried lamb. Convert it to yield 150 portions.

Ingredients for 50 Portions	Amount of Conversion	Amount Needed to Yield 150 Portions
18 lb. lamb shoulder, boneless, cut into 1-inch cubes		
2 ½ gal. water		
2 lb. butter or shortening		
1 lb. 8 oz. flour		
1/3 cup curry powder		
2 qt. tart apples, diced		
2 lb. onions, diced		
½ tsp. ground cloves		
½ tsp. nutmeg		
2 bay leaves		
1 tsp. marjoram		
Salt and pepper to taste		

2. The following recipe yields 50 portions of veal paprika with sauerkraut. Convert it to yield 40 portions.

Ingredients for 50 Portions	Amount of Conversion	Amount Needed to Yield 40 Portions
15 lb. veal shoulder, cut into 1-inch cubes		
1 lb. butter		
6 lb. onions sliced thin		
10 lb. sauerkraut with juice		
3 tbsp. salt		
5 tbsp. paprika		
1 tbsp. pepper, fresh ground		
1 qt. sour cream, thick		
3 cloves garlic, minced		

3. The following recipe yields 100 portions of Hungarian goulash. Convert it to yield 75 portions.

Ingredients for 100 Portions	Amount of Conversion	Amount Needed to Yield 75 Portions
36 lb. beef chuck or shoulder, diced 1-inch cubes		
1 ½ oz. garlic, minced		
1 lb. 4 oz. flour		
1 ¼ oz. chili powder		
10 oz. paprika		
2 lb. tomato puree		
2 gal. brown stock		
4 bay leaves		
¾ oz. caraway seeds		
3 lb. 8 oz. onions, minced		
salt and pepper to taste		

4. The following recipe yields 50 portions of salisbury steak. Convert it to yield 425 portions.

Ingredients for 50 Portions	Amount of Conversion	Amount Needed to Yield 425 Portions
14 lb. boneless beef chuck		
3 lb. onions, minced		
1 tsp. garlic, minced		
½ cup salad oil		
8 whole eggs		
2 lb. bread cubed		
1 ½ pt. milk		
salt and fresh ground pepper to taste		

5. The following recipe yields nine 8-inch lemon pies. Convert it to yield six 8-inch pies.

Ingredients for 9 Pies	Amount of Conversion	Amount Needed to Yield 6 Pies
4 lb. water		
3 lb. 6 oz. granulated sugar		
½ oz. salt		
3 oz. lemon gratings		
1 lb. water		
8 oz. corn starch		
12 oz. egg yolks		
1 lb. 6 oz. lemon juice		
4 oz. butter, melted		
yellow color, as needed		

6. The following recipe yields nine 8-inch lemon chiffon pies. Convert it to yield thirty-six 8-inch pies.

Ingredients for 9 Pies	Amount of Conversion	Amount Needed to Yield 36 Pies
3 lb. water		
2 lb. granulated sugar		
¾ oz. salt		
2 oz. lemon grating (rind)		
1 lb. egg yolks		
1 lb. lemon juice		
9 oz. cornstarch		
1 ½ oz. plain gelatin		
1 lb. hot water		
2 lb. egg whites		
1 lb. 8 oz. granulated sugar		
yellow color as needed		

7. The following recipe yields 12 dozen hard rolls. Convert it to yield 9 dozen rolls.

Ingredients for 12 Doz. Rolls	Amount of Conversion	Amount Needed to Yield 9 Doz. Rolls
7 lb. 8 oz. bread flour		
3 oz. salt		
3 ½ oz. granulated sugar		
3 oz. shortening		
3 oz. egg whites		
4 lb. 8 oz. water (variable)		
4 ½ oz. yeast, compressed		

8. The following recipe yields 12 dozen soft rye rolls. Convert it to yield 78 dozen rolls.

Ingredients for 12 Doz. Rolls	Amount of Conversion	Amount Needed to Yield 78 Doz. Rolls
6 lb. 6 oz. bread flour		
1 lb. 4 oz. rye flour dark		
6 oz. yeast, compressed		
1 ¾ oz. salt		
5 oz. dry milk		
1 lb. shortening (hydrogenated)		
1 lb. sugar		
1 ½ oz. malt		
4 lb. 8 oz. water (variable)		
6 oz. caraway seeds		

9. The following recipe yields 8 dozen soft dinner rolls. Convert it to yield 5 dozen rolls.

Ingredients for 8 Doz. Rolls	Amount of Conversion	Amount Needed to Yield 5 Doz. Rolls
10 oz. granualted sugar		
10 oz. hydrogenated shortening		
1 oz. salt		
3 oz. dry milk		
4 oz. whole eggs		
3 lb. 12 oz. bread flour		
2 lb. water		
5 oz. yeast, compressed		

10. The following recipe yields 9 dozen cornmeal muffins. Convert it to yield 45 dozen muffins.

Ingredients for 9 Doz. Muffins	Amount of Conversion	Amount Needed to Yield 45 Doz. Muffins
2 lb. 8 oz. granulated sugar		
1 ½ oz. salt		
6 oz. powder milk		
1 lb. 8 oz. whole eggs		
2 lb. water		
3 lb. 12 oz. bread flour		
1 lb. 8 oz. cornmeal		
5 oz. baking powder		
1 lb. water		
1 lb. 8 oz. salad oil		

UNIT 7
Daily Food Production Reports

Food production reports are a necessary part of a food service operation, figure 7-1. They are probably more important in large operations because they help the manager or chef control production situations that can become out of control. In small operations with a smaller work crew, less production, and a smaller physical work area, it is easier to observe all that is taking place, so production reports are not as important. Large catering companies that have many units throughout a city or state rely heavily on these reports. The purpose of daily food production reports can be expressed in one word: control. They control:

- over or under production
- leftovers
- purchasing
- labor cost
- waste
- theft

This type of report is also used in predicting future sales, sometimes referred to as *forecasting*. Predicting future sales assists management in the purchasing of food and hiring of future food service employees. If a rotating menu is being used on a monthly, six months, or yearly basis, these reports become even more valuable. Management can check back and see which items sold best and how many of each item was sold the last time the menu was used. If daily food production reports are used properly, management can improve the food cost percentage and this, of course, is the major goal.

The forms used in compiling the daily food production reports vary depending on the individual food operation. Each establishment has

Fig. 7-1 Cook completing a daily production report.

its own ideas about control. Most establishments do, however, request reports from the cooks, pastry cooks, salad department, and counter service. The report forms are usually quite simple to fill out so they do not take up too much of an employee's time to complete. Some forms, such as the one used in the counter report, show unit price, total price, and total sales.

Examples of four different daily food production reports are given in figures 7-2 through 7-5. The forms shown are typical of most operations.

Although the examples of daily food production reports shown are easy to follow, a few comments on each can help the student become more competent when making them out.

Cook's Production Report

Recipe File Number. This number is placed on the recipe when it is filed. Standard recipes

are used in most food service establishments to control cost, taste, texture, quality, and amounts of food being prepared. The manager or chef usually fills in the number to indicate to the cook the recipe to be used.

Size of Portion. The manager or chef usually fills in this column to let the cook know the portion size to use when serving. In most establishments, the portion size is a set policy and is indicated on the portion charts on display in the production area.

Raw Quantity Required. This is usually designated by the chef or cook. Sometimes the manager lists this figure, but it must be done by a person familiar with production and with the policy of the establishment.

Portions to Prepare. This decision is made by the manager or chef and is based on previous

Unit — First National Bank		Meal — Luncheon		Customers — 150		
Day — Thursday		Date — June 23, 19__				
Item	Recipe File Number	Size of Portion	Raw Quantity Required	Portions to Prepare	Portions Left or Time Out	Portions Served
Roast Round of Beef	15	3 oz.	25 lb.	80	out 8:15 p.m.	80
Roast Loin of Pork	20	4 ½ oz.	12 lb.	30	8	22
Filet of Sole	8	5 oz.	14 lb.	42	2	40
Veal Goulash	18	6 oz.	10 lb.	24	9	15
Swiss Steak	14	5 oz.	15 lb.	48	12	36
Mashed Potatoes	42	4 oz.	20 lb.	60	5	55
Peas and Carrots	51	3 oz.	9 lb.	48	18	30
Succotash	52	3 oz.	5 lb.	26	1	25

Fig. 7-2 Cook's production report.

Day — Monday		Date — June 27, ___			Unit — Leo's Cafeteria	
Item	**Order**	**On Hand**	**Prepare**	**Left**	**Sold**	**Comments**
Rolls						
Soft Rye	20 doz.	3 doz.	17 doz.	2 doz.	18 doz.	
Soft White	35 doz.	4 doz.	31 doz.	6 doz.	29 doz.	
Hard White	26 doz.	5 doz.	21 doz.	4 doz.	22 doz.	
Cinnamon	18 doz.	2 doz.	16 doz.	1 doz.	17 doz.	
Rye Sticks	15 doz.	6 doz.	9 doz.	0	15 doz.	Out 7 p.m.
Quick Breads						
Biscuits	12 doz.	1 doz.	11 doz.	3 doz.	9 doz.	Biscuits left
Raisin Muffins	24 doz.	5 doz.	19 doz.	5 doz.	19 doz.	unbaked in freezer
Pies						
Cherry	22	6	16	0	22	Out 8 p.m.
Apple	25	4	21	2	23	
Chocolate	15	3	12	0	15	Out 7:30
Banana	12	1	11	8	4	p.m.
Cakes						
Bar	6	2	4	3	3	These left
8″ White	10	4	6	6	4	in freezer
Mocha	8	1	7	0	8	Out 7 p.m.

Fig. 7-3 Bakery production report.

production reports. How much of a particular item was sold the last time it appeared on the menu is very important. Weather conditions the last time also influence this decision.

Portions Left or Time Out. This figure is recorded by the cook and is found by counting the number of portions left, figure 7-6, after the meal (luncheon, dinner, etc.) is over. Time out is recorded when the number of a certain item is completely sold out. The time an item is sold out is important because it affects the number of items prepared the next time the item appears on the menu.

Portions Served. This is recorded by the cook and is found by subtracting the number left from the number prepared. It is an important figure because this number influences the number of portions to prepare when the item appears again on the menu.

Bakery Production Report

Order. Recorded by the manager, chef, or pastry chef, this represents the amount needed for service throughout the day in all departments of the establishment. If the establishment is a

Day — Friday		Date — June 29, 19__		Unit — First National Bank	
Item	**Order**	**On Hand**	**Prepare**	**Left**	**Sold**
Toss	23	8	15	3	20
Italian	12	0	12	0	12
Garden	22	9	13	7	15
Waldorf	16	2	14	6	10
Potato	25	10	15	6	19
Cole Slaw	32	8	24	1	31
Sl. Tomato	26	0	26	3	23
Fruited Gelatin	28	12	16	2	26
Sunshine	14	3	11	4	10
Green Island	25	5	20	5	20
Mixed Fruit	35	6	29	3	32
Chef	46	7	39	12	34
Cucumber	12	1	11	10	2

Fig. 7-4 Salad production report.

Unit — Latonia Race Track			Customer Count — 405		
Day — Tuesday			Date — June 28, 19__		
Item	**Number of Portions For Sale**	**Number of Portions Not Sold**	**Number of Portions Sold**	**Unit Price**	**Value Sold**
Hot Dogs	135	26	109	$ .55	$59.95
Metts	75	15	60	.65	39.00
Hamburgers	150	23	127	.75	95.25
Barbecue	50	5	45	.60	27.00
Cube Steaks	70	6	64	1.25	80.00
Milk	125	18	107	.25	26.75
Shakes	80	7	73	.55	40.15
Soda	225	28	197	.25	49.25
Cake	15	2	13	.45	5.85
Pie	35	8	27	.35	9.45
Ice Cream	65	9	56	.25	14.00
Potato Chips	85	13	72	.20	14.40
Pretzels	45	11	34	.20	6.80
Name Bill Thompson				**Total**	$467.85

Fig. 7-5 Counter production report.

Fig. 7-6 Cook completing cook's production report by counting the leftover portions.

catering company, it represents the amount needed in all units served.

On Hand This figure is recorded by the baker and represents the amount of each item that was left by the previous shift or from the previous day and is still in a usable condition. It may be in a raw or cooked state, frozen or unfrozen.

Prepare. The number to prepare is found by subtracting the amount on hand from the amount ordered. The remainder is the amount to prepare and is recorded by the baker or pastry chef.

Left. The number left is found by counting the number of pieces remaining, figure 7-7, after the day's service is over. This figure is recorded by the baker or pastry chef.

Sold. The number sold is found by subtracting the number left from the number ordered. This figure is recorded by the cook or pastry chef.

Comments. This space is provided for any information that may be valuable to management. Examples: The time a product is sold out; the disposition of leftover items.

Fig. 7-7 Cook completing the pantry production report by counting the number of portions left after service.

Salad Production Report

Order. Recorded by the manager, chef, or head salad person, this represents the amount of each salad to be prepared by the salad person. It is an estimate of the number of each salad needed for serving one meal or for the complete day.

On Hand. This figure is recorded by the salad person and represents the amount of each salad that was left by the previous shift or from the previous day and is still in a usable condition.

Prepare. The number to prepare is found by subtracting the amount on hand from the amount ordered. This figure is recorded by the salad person.

Left. The number left is found by counting the number of each kind of salad remaining, figure 7-8, after the meal or day's service is concluded. The salad person records this figure.

Sold. The number sold is found by subtracting the number left from the number ordered. This figure is recorded by the salad person.

Comments. This space is provided for information that may be valuable to management.

Examples: The time a certain salad is sold out; production information or mistakes; and the disposition of leftover salads.

Counter Production Report

Number of Portions for Sale. This figure may be recorded by management or the person working

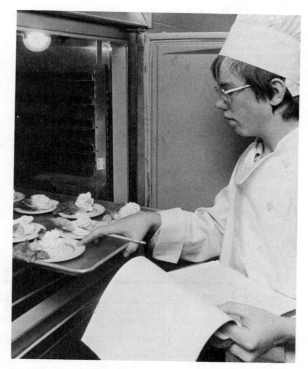

Fig. 7-8 Cook completing a salad production report by counting the number of portions remaining after service.

the counter. It represents the number of each on-hand item that is for sale.

Number of Portions Not Sold. This figure is recorded by the person working the counter. It is found by counting the remaining pieces of each item left after the day's service is concluded.

Number of Portions Sold. This is found by subtracting the number of portions not sold from the number of portions for sale. It is recorded by the person working the counter.

Unit Price. The cost of one item. It is determined and recorded by the manager.

Value Sold. This is found by multiplying the number of portions sold by the unit price. It is found and recorded by the person working the counter.

Total. This is found by adding the figures in the value sold column. It is found and recorded by the person working the counter.

Customer Count. This figure is recorded on the register. The person working the counter should check the register and record this figure right before going off duty.

ACHIEVEMENT REVIEW
PRODUCTION REPORTS

A. Make up cook's production reports using the information given below. Use the same type of form shown in figure 7-2, with the present day and date. Complete the reports.

1.

Unit — Mason Art Co.		Meal — Luncheon			Customers — 60	
Day —		Date —				

Item	Recipe Number	Size of Portion	Raw Quantity Required	Portions to Prepare	Portions Left or Time Out	Portions Served
Spanish Steak	18	5 oz.	8 lb.	25	7	
Salisbury Steak	19	5 oz.	7 lb.	22	9	
Beef Pot Roast	20	3 oz.	8 lb.	32	13	
Au Gratin Potatoes	2	4 oz.	10 lb.	35	6	
KY Succotash	10	3 oz.	5 lb.	30	4	
Mashed Potatoes	3	4 oz.	20 lb.	60	13	
Lima Beans	12	3 oz.	2 ½ lb.	15	2	

2.

Unit — 2nd National Bank	Meal — Luncheon			Customers — 103		
Day —			Date —			

Item	Recipe Number	Size of Portion	Raw Quantity Required	Portions to Prepare	Portions Left or Time Out	Portions Served
Sautéed Pork Chop	32	4 oz.	6 lb.	22	8	
Turkey Steaks	45	4 oz.	12 lb.	45	7	
Baked Halibut	54	5 oz.	10 lb.	32	12	
Roast Veal	30	3 oz.	10 lb.	38	13	
Parsley Potatoes	4	4 oz.	15 lb.	50	6	
Hash in Cream Potatoes	5	4 oz.	14 lb.	45	11	
Peas and Celery	13	3 oz.	7 ½ lb.	44	9	
Cut Green Beans	16	3 oz.	5 lb.	30	5	

3.

Unit — Chase Machine Tool Co.	Meal — Luncheon			Customers — 158		
Day —			Date —			

Item	Recipe Number	Size of Portion	Raw Quantity Required	Portions to Prepare	Portions Left or Time Out	Portions Served
Sautéed Veal Steak	31	4 oz.	5 lb.	26	3	
Beef Sauerbraten	21	3 oz.	8 lb.	32	6	
Beef Goulash	26	6 oz.	15 lb.	54	2	
Swiss Steak	22	5 oz.	20 lb.	62	8	
Rissel Potatoes	1	4 oz.	12 lb.	36	9	
Escallop Potatoes	6	4 oz.	8 lb.	26	11	
Stewed Tomatoes	14	3 oz.	1 #10 can	22	5	
Corn	15	3 oz.	7 ½ lb.	44	12	

4.

Unit — Norwood High School		Meal — Luncheon			Customers — 170	
Day —		Date —				

Item	Recipe Number	Size of Portion	Raw Quantity Required	Portions to Prepare	Portions Left or Time Out	Portions Served
Ham Steak	34	4 oz.	7 lb.	28	6	
Roast Turkey	37	3 oz.	15 lb.	36	8	
Hamburger Steak	23	5 oz.	13 lb.	42	7	
Beef Stroganoff	25	6 oz.	7 lb.	20	5	
Macaroni Au Gratin	74	4 oz.	4 lb.	45	11	
Mashed Potatoes	3	4 oz.	12 lb.	40	9	
Mixed Greens	9	3 oz.	10 lb.	58	13	
Carrots Vichy	17	3 oz.	5 lb.	29	16	

5.

Unit — Harold Steel Mill		Meal — Luncheon			Customers — 110	
Day —		Date —				

Item	Recipe Number	Size of Portion	Raw Quantity Required	Portions to Prepare	Portions Left or Time Out	Portions Served
Roast Breast of Chicken	45	4 oz.	15 lb.	56	4	
Broiled Pork Chop	35	5 oz.	8 lb.	25	9	
Sautéed Veal Chop	33	5 oz.	12 lb.	36	8	
Broiled Flank Steak	24	4 oz.	6 lb.	22	13	
Duchess Potatoes	7	4 oz.	10 lb.	38	18	
Croquette Potatoes	8	3 oz.	7 lb.	32	19	
Corn Obrien	11	3 oz.	5 lb.	35	9	
Peas and Onions	13	3 oz.	7 ½ lb.	45	12	

B. Make up bakery production reports using the information given below. Use the same type of form shown in figure 7-3, with present day and date. Complete the reports.

1.

Day —	Date —		Unit — Norwood High School	
Item	Order	On Hand	Left	Sold
Seed Rolls	25 doz.	6 doz.	4 doz.	
Rye Rolls	36 doz.	3 doz.	5 doz.	
Hard Rolls	28 doz.	6 doz.	2 doz.	
Rye Sticks	22 doz.	4 doz.	3 doz.	
Biscuits	12 doz.	0	3 doz.	
Banana Muffins	10 doz.	0	1 ½ doz.	
Cherry Pies	18	5	2	
Banana Pies	16	2	4	
Boston Cream Pies	22	7	6	
Deviled Food Cake	13	5	2	

2.

Day —	Date —		Unit — Deluxe Shoe Co.	
Item	Order	On Hand	Left	Sold
Soft Rolls	30 doz.	4 doz.	2 doz.	
Hard Rolls	28 doz.	6 doz.	5 doz.	
Rye Sticks	22 doz.	4 doz.	1 doz.	
Pecan Rolls	16 doz.	2 doz.	½ doz.	
Raisin Muffins	9 doz.	1 doz.	¾ doz.	
Apple Pies	18	3	1	
Peach Pies	14	2	2	
Chocolate Pies	22	4	3	
Lemon Cake	9	2	4	
Chocolate Bar Cake	12	5	3	
Fudge Cake	14	1	5	

3.

Day —	Date —		Unit — Joe's Cafeteria	
Item	Order	On Hand	Left	Sold
Cloverleaf Rolls	40 doz.	7 doz.	3 doz.	
Caramel Rolls	26 doz.	3 doz.	1 doz.	
Seed Rolls	35 doz.	4 doz.	2 doz.	
Biscuits	22 doz.	5 doz.	½ doz.	
Corn Muffins	18 doz.	2 doz.	¾ doz.	
Pecan Pie	12	3	5	
Pumpkin Pie	15	2	4	
Custard Pie	10	4	3	
Eclair	48	0	8	
Cherry Tarts	54	5	7	

4.

Day —	Date —		Unit — Wine & Dine Restaurant	
Item	Order	On Hand	Left	Sold
Soft Rolls	68 doz.	9 doz.	4 ¾ doz.	
Rye Rolls	38 doz.	4 doz.	5 ½ doz.	
Split Rolls	26 doz.	7 doz.	2 ¼ doz.	
Cinnamon Rolls	32 doz.	6 doz.	7 ¾ doz.	
Apple Muffins	18 doz.	4 doz.	4 ½ doz.	
Corn Sticks	16 doz.	3 doz.	2 ¼ doz.	
Blueberry Pie	12	4	2	
Coconut Cream Pie	14	3	4	
White Cake, 8"	9	2	1	
Apple Cake, 8"	8	1	0	
Yellow Cake, 8"	6	0	3	

5.

Day —	Date —		Unit — United Shoe Co.	
Item	Order	On Hand	Left	Sold
Soft Rolls	48 doz.	6 doz.	2 doz.	
Rye Rolls	32 doz.	5 doz.	3 doz.	
Rye Sticks	28 doz.	3 doz.	2 ½ doz.	
Cinnamon Rolls	22 doz.	2 doz.	1 ¾ doz.	
Corn Muffins	18 doz.	3 doz.	6 ¾ doz.	
Bran Muffins	15 doz.	4 doz.	2 doz.	
Apple Pies	14	3	1	
Peach Pies	18	6	2	
Lemon Meringue Pies	20	5	0	
Yellow Cake, 8″	10	1	3	
White Cake, 8″	8	2	2	
Peach Bar Cakes	6	0	2	

C.   Make up salad production reports using the information given below. Use the same type of form shown in figure 7-4, with the present day and date. Complete the reports.

1.

Day —	Date —		Unit — 1st Federal Bank	
Item	Order	On Hand	Left	Sold
Toss	24	4	4	
Chef	20	3	2	
Garden	16	6	6	
Slaw	12	2	3	
Gelatin	9	1	7	
Fruit	15	9	5	
Sl. Tomato	18	4	1	
Waldorf	8	5	0	

2.

Day —		Date —		Unit — 2nd Federal Bank
Item	**Order**	**On Hand**	**Left**	**Sold**
Italian	22	5	1	
Cucumber	24	4	4	
Jellied Slaw	18	2	5	
Green Island	16	1	2	
Macaroni	12	3	6	
Carrots	15	6	3	
Mixed Green	14	4	0	
Sunshine	26	2	1	

3.

Day —		Date —		Unit — Western Insurance Co.
Item	**Order**	**On Hand**	**Left**	**Sold**
Chef	32	6	7	
Waldorf	18	7	6	
Fruited Slaw	14	8	5	
Fruited Gelatin	16	4	1	
Sl. Tomato	25	3	0	
Garden	28	2	2	
Italian	35	0	3	
Mixed Green	40	1	4	

4.

Day —		Date —		Unit — Garrison Greeting Card Co.
Item	**Order**	**On Hand**	**Left**	**Sold**
Toss	35	4	0	
Mixed Green	25	8	2	
Garden	20	6	4	
Spring	22	2	7	
Cottage Cheese	18	0	6	
Waldorf	16	3	3	
Sunshine	14	0	2	
Hawaiian	12	1	0	
Macaroni	10	7	1	

5.

Day —	Date —		Unit — Link Machine Tool Co.	
Item	Order	On Hand	Left	Sold
Mixed Green	16	1	0	
Chef	28	4	4	
Western	30	5	2	
Italian	16	3	6	
Green Island	12	6	1	
Fruited Gelatin	10	2	3	
Cole Slaw Soufflé	14	0	5	
Waldorf	20	5	4	

D. Make up counter production reports using the information given below. Use the same type of form shown in figure 7-5, with present day and date. Complete the reports.

1.

Unit — Stevens Processing Co.		Customer Count — 305		
Day —		Date —		
Item	Number of Por-tions for Sale	Number of Portions Sold	Unit Price	Value Sold
Hot Dogs	85	80	$ .55	
Hamburgers	95	75	.75	
Metts	65	63	.65	
Barbecue	55	53	.60	
Cube Steak	40	32	1.25	
Soda	120	109	.25	
Shakes	60	56	.55	
Milk	110	107	.25	
Pie	48	42	.35	
Ice Cream	60	38	.25	

2.

| Unit — Wall Manufacturing Plant | | Customer Count — 435 | | |
| Day — | | Date — | | |
Item	Number of Por- tions for Sale	Number of Portions Sold	Unit Price	Value Sold
Hot Dogs	120	89	$ .55	
Hamburgers	115	106	.75	
Metts	112	98	.65	
Barbecue	75	58	.60	
Cube Steaks	85	81	1.25	
Soda	125	106	.25	
Shakes	90	79	.55	
Milk	120	99	.25	
Pie	90	83	.35	
Ice Cream	90	82	.25	

3.

| Unit — Deluxe Playing Card Co. | | Customer Count — 308 | | |
| Day — | | Date — | | |
Item	Number of Por- tions for Sale	Number of Portions Sold	Unit Price	Value Sold
Hamburgers	90	82	$ .75	
Hot Dogs	85	73	.55	
Metts	70	68	.65	
Barbecue	50	43	.60	
Cube Steaks	45	38	1.25	
Soda	95	87	.25	
Shakes	45	41	.55	
Milk	85	76	.25	
Pie	54	39	.35	
Cake	60	42	.30	

4.

| Unit — United Shoe Co. | | Customer Count — 390 | | |
| Day — | | Date — | | |
Item	Number of Portions for Sale	Number of Portions Sold	Unit Price	Value Sold
Hamburgers	110	102	$ .75	
Hot Dogs	105	96	.55	
Metts	80	74	.65	
Barbecue	75	66	.60	
Cube Steaks	65	52	1.25	
Soda	100	89	.25	
Shakes	85	72	.55	
Milk	95	69	.25	
Pie	66	58	.35	
Cake	40	36	.30	

5.

| Unit — Reeves Tire Plant | | Customer Count — 285 | | |
| Day — | | Date — | | |
Item	Number of Portions for Sale	Number of Portions Sold	Unit Price	Value Sold
Hamburgers	75	69	$ .75	
Hot Dogs	85	74	.55	
Metts	70	49	.65	
Barbecue	60	59	.60	
Cube Steak	40	32	1.25	
Soda	85	82	.25	
Shakes	70	67	.55	
Milk	80	75	.25	
Pie	42	39	.35	
Cake	48	43	.30	

UNIT 8
Back-of-the-House Business Forms

All business ventures involve some paperwork. Records must be kept of all the business that is carried on within an organization. Records need not be complicated. They can be quite simple to complete, so that they do not take up too much time. Behind every successful business is usually a good record keeper.

Business forms used in food service operations vary depending on the accounting system. An accounting system is usually set up by a public accountant specializing in accounting for restaurants, who is hired on a part-time basis by the restaurant. The accountant usually provides forms that must be kept up to date and which reflect the daily business operation. It is important to keep a daily record of such items as the cash register readings, cash on hand, bank deposits, cash paid-outs, and checks issued. These forms are handled by management and office personnel.

In addition to the forms mentioned, there are at least four other important business forms that the production worker comes in contact with: requisitions, invoices, purchase orders, and inventories. These are an important part of any business operation that buys and sells a product, and everyone working in a restaurant or food service establishment should be familiar with them. Inventories will be discussed in another unit. This unit deals with the other three forms.

REQUISITIONS

A cook in the kitchen needs certain supplies to carry out the day's production. All supplies are stored in the storeroom. To obtain these supplies, the cook fills out a *storeroom requisition* such as the one in figure 8-1. On the form, the cook states the quantity needed, the unit, and a description of each item. The requisition must be approved by a superior. The cook who is going to use the items must also sign the requisition. This requisition is then taken to the storeroom, figure 8-2, and the supplies are issued. The person in charge of the storeroom, usually referred to as the *steward,* marks the unit price and extension price on each item. The steward then finds the total price of the food issued. The requisition is used for the following purposes:

- Account for all items issued from the storeroom each day.
- Control theft and waste.

Storeroom Requisition

Date October 23, 19 _____   Charge To Kitchen

Quantity	Unit	Item	Unit Price	Extension Price
6	#10 cans	Sliced Apples	$3.50	$21.00
4	#10 cans	Tomato Juice	1.95	7.80
1	lb.	Fresh Mushrooms	1.85	1.85
2	lb.	Beef Base	2.20	4.40
4	lb.	Cornstarch	.69	2.76
			Total	$37.81

Approved by *Robert Haines*
Signed *John Doe*

Fig. 8-1 Requisition form.

- Provide the figures necessary for the daily food cost report.
- Ensure that all items are issued only to the proper personnel.
- Assist in controlling purchasing and eliminating large inventories.

INVOICES

An *invoice,* figure 8-3, is a written document listing goods sent to a purchaser with their prices, quantity, and charges. Individual companies have their own type of invoices depending upon what they feel is necessary to list. Some invoices are simple, others more complex.

An invoice accompanies each shipment or delivery of food that is brought into a restaurant. Before signing for a shipment, the person receiving the delivery should check the items delivered with those listed on the invoice to make sure all items have been received. The invoice is very important to the restaurant operator or manager because it provides the figures for the food purchased during the month. These figures are necessary when computing the monthly food cost percent. The invoice is also important when checking the charges listed on the monthly bill or statement sent by the vendor. Some business people pay on the invoice, but most wait for a monthly bill or statement and check all charges

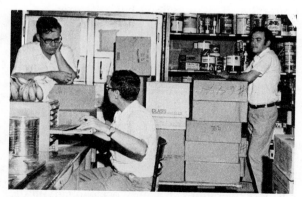

Fig. 8-2 Requisitions sent by cooks for food service supplies are filed in the storeroom.

Distributor:	Haines Foods, Inc.			Phone: _____	
Address:	70 Greenbrier Ave.				
	Ft. Mitchell, KY 41017				

Distributors of Fine Food Products

Wholesale Only Date: <u>October 20, 19</u> _____

No. of Pieces	Salesperson	Order No.	Invoice No.
5	Joe Jones	2860	J 2479

Packed By
 G.C.

Sold To: Mr. John Doe
Street: 120 Elm Ave.
City: Covington, KY

Case	Pack	Size	Canned Foods	Price	Amount
4	6	#10 can	Sliced Apples	$ 8.95	$ 35.80
3	6	#10 can	Pitted Cherries	15.60	46.80
2	12	#5 can	Apple Juice	9.50	19.00
1	24	1 lb.	Cornstarch	8.40	8.40
2	24	#2 1/2 can	Asparagus	18.90	37.80
				Total Amount	$147.80

Fig. 8-3 Invoice form.

against the invoices received with each delivery, figure 8-4. This is a good business practice.

If an establishment has a policy of using purchase orders, then the invoice can be used by the bookkeeper to check against a copy of the purchase order before the bill is paid.

PURCHASE ORDERS

Purchase orders are used often in certain business operations, but only occasionally in the food service business. Large restaurant chains are more likely to use them than small or independent operators.

A *purchase order,* figure 8-5, is a written form which indicates to the vendor how many items are to be delivered to an establishment, and lists the prices for each item. The order is usually signed by the owner, manager, purchasing agent, or chef. This form tells the vendor that if the items delivered compare favorably with the items and prices listed on the order, payment will be made.

In most food service operations, one person is designated to do the purchasing. This person

Fig. 8-4 The weight of incoming meats should be checked against the amount stated on the invoice.

is usually called the *purchasing agent*. In small operations, this duty is often performed by the manager, assistant manager, or chef. In any case, the person doing the buying checks prices and quality with about three different vendors before the purchase order is sent to the one offering the best deal.

The purchase order is usually made out in triplicate (3 copies). The original copy is sent to the vendor and carbon copies are kept by the steward and the bookkeeper. The steward uses a copy of the purchase order to check the mer-

chandise, figure 8-6, when it is delivered by the vendor. The bookkeeper uses a copy to check what was ordered against invoices and statements. Purchase orders eliminate controversy over what was ordered, how much was ordered, who ordered it, and when it was ordered.

PURCHASE SPECIFICATIONS

Purchase specifications are an important part of a successful food service operation. They

Hasenours Restaurant Barret and Oak Sts. Louisville, KY	Purchase Order No. 1492 Date: January 6, 19 ____

To: Jefferson Meat Co.
     2868 Baxter Ave.
     Louisville, KY

Ship To: Hasenours Restaurant
          Barret and Oak Sts.
          Louisville, KY

Date of Delivery: January 10, 19 ____

   Deliver the items listed below, which are being purchased in accordance with descriptions and prices stated.

Description	Unit	Quantity	Unit Price	Amount
Ribs of Beef—Choice 20 to 22 lb. Aged 15 to 20 days Short Ribs removed	lb.	132 lb.	$2.95	$ 389.40
12 oz. Sirloin Steaks—Choice 1 1/2 inch thick Packed for storage	lb.	288 lb.	$3.75	$1080.00
Beef Chuck for Stew Cut into 1-inch cubes Grade Choice	lb.	40 lb.	$2.28	$ 91.20
			Total Cost	$1560.60

Robert Haines
Purchasing Agent

Fig. 8-5 Purchase order form.

provide a detailed description of the items being purchased.

Most restaurants use purchase specifications when purchasing meats, seafoods, and produce (fruits and vegetables). For example, when purchasing ribs of beef, the specifications may be:

Fig. 8-6 All incoming supplies must be checked against those listed on the invoice.

1. Grade—Choice

2. Weight—20 to 22 lb.

3. Short ribs removed after measuring 1½ inches from the "eye" of the rib.

4. The back should not have a heavy covering of fat.

5. Ribs should be aged 15 to 20 days.

6. Back bones should be separated from the seven rib bones.

7. Rib tied—Oven ready

When purchasing fruit, the specifications usually list the size, weight, softness, brand, number desired, and when appropriate, the color. A copy of purchase specifications is sent to the vendor or discussed with the vendor before the food service establishment begins buying. In some cases, however, the purchase specifications are sent with the purchase order or written on the purchase order.

ACHIEVEMENT REVIEW
BACK-OF-THE-HOUSE BUSINESS FORMS

A. Prepare 5 storeroom requisition forms as shown in figure 8-1. Work the following problems by finding the extension price and the total.

1. 8–#10 cans sliced peaches @ $1.96 per can
   7–#10 cans whole tomatoes @ $1.23 per can
   12 heads of iceburg lettuce @ $.49 per head
   3 dozen eating apples @ $.98 per dozen
   5 pounds clear jell starch @ $.43 per pound
   4 pounds margarine @ $.59 per pound

2. 5–#10 cans sliced pineapple @ $1.89 per can
   4–#10 cans cherries @ $2.30 per can
   3 bunches celery @ $.49 per bunch
   6 pounds tomatoes @ $.45 per pound
   2 bunches carrots @ $.22 per bunch
   9 dozen eggs @ $.79 per dozen

3. 8–13-ounce cans tuna fish @ $1.36 per can
   5 dozen eggs @ $.79 per dozen
   9–1-pound cans salmon @ $1.29 per can
   6–2½-pound boxes frozen peas @ $.36 per pound
   8–2½-pound boxes frozen corn @ $.34 per pound
   7 heads lettuce @ $.49 per head

4. 3–#10 cans tomato puree @ $1.12 per can
   4–1-pound boxes cornstarch @ $.69 per pound
   2½ pounds leaf lettuce @ $.70 per pound
   5–2½-pound boxes lima beans @ $.38 per pound
   2½ pounds fresh mushrooms @ $1.45 per pound
   4 bunches green onions @ $.23 per bunch
   3 bunches parsley @ $.21 per bunch
   5 heads lettuce @ $.49 per head

5. 9–1-pound cans asparagus spears @ $.89 per can
   6–5-pound boxes frozen mixed vegetables @ $.36 per pound
   3½ pounds Romaine lettuce @ $.84 per pound
   2¾ pounds Endive lettuce @ $.75 per pound
   2 pounds salt @ $.16 per pound
   8 pounds butter @ $.88 per pound
   5 gallons salad oil @ $2.23 per gallon
   20 pounds Idaho potatoes @ $.24 per pound

B. Prepare 5 invoice forms as shown in figure 8-3. Work the following problems by finding the extension price and the total. Assume that you work for Haines Foods, Inc. Use today's date and your own name as salesperson.

1. Order No. 2861; Invoice No. J 2480; Packed by R.H.; Sold to Manor Restaurant, 590 Walnut Street, Cincinnati, OH

4 cases 6–#10 cans whole tomatoes @ $5.60 per case
6 cases 12–#5 cans tomato soup @ $7.80 per case
2 cases 20 pounds spaghetti @ $4.95 per case
4½ cases 4–1-gallon jars mayonnaise @ $16.80 per case
1½ cases 6–#10 cans bacon bits @ $12.64 per case
2 cases 12–2-pound cans regular coffee @ $48.00 per case

2. Order No. 2862; Invoice No. J2481; Packed by R.H.; Sold to Sinton Hotel, 278 Vine St., Cincinnati, OH

3 cases 4–1-gallon whole dill pickles @ $7.28 per case
6 cases 12–#5 cans chicken rice soup @ $18.70 per case
4 cases 4–1-gallon sweet relish @ $9.40 per case
6 cases 6–#10 cans pumpkin @ $7.34 per case
5 cases 6–#10 cans sliced pineapple @ $15.60 per case
½ case 2–1-gallon red maraschino cherries @ $19.20 per case

3. Order No. 2863; Invoice No. J2482: Packed by R.H.; Sold to Cincinnati Businessmen's Club, 529 Plum St., Cincinnati, OH

3 cases 6–#10 cans catsup @ $16.34 per case
5 cases 24–1-pint bottles chili sauce @ $15.60 per case
2 cases 4–1-gallon cider vinegar @ $4.48 per case
2 cases 20 pounds lasagne @ $2.29 per case
9 cases 6–#10 cans sliced peaches @ $14.10 per case
1 pkg. 1-pound black pepper @ $1.08 per pound

4. Order No. 2864; Invoice No. J2483; Packed by R.H.; Sold to Norwood High School Cafeteria, 2078 Elm Ave., Norwood, OH

7 cases 6–#10 cans apple slices @ $9.40 per case
4 cases 10 pounds medium egg noodles @ $3.20 per case
2 cases 6–#10 cans bean sprouts @ $5.10 per case
2 cases 32 pounds margarine @ $15.36 per case
6 cases 24–1-pound boxes brown sugar @ $6.72 per case
3 bags 100 pounds granulated sugar @ $15.00 per bag

5. Order No. 2865; Invoice No. J2484; Packed by R.H.; Sold to Scarlet Oaks Career Development Center, 3254 East Kemper Rd., Cincinnati, OH

2 cases 4–1-gallon Worcestershire sauce @ $6.60 per case
12 cases 6–#10 cans cut green beans @ $6.22 per case
8 cases 12–#5 cans tomato juice @ $2.45 per case
9 cases 12–#5 cans apple juice @ $2.65 per case
2 cases 12–5-pound boxes frozen cod fish @ $53.40 per case
2 slabs 8-pound frozen white meat turkey @ $1.75 per pound
2 cases 4–1-gallon prepared mustard @ $5.00 per case
11 cases 6–#10 cans sliced apples @ $9.40 per case

C. Prepare 5 purchase order forms as shown in figure 8-5. Work the following problems by finding the amount and total cost for all items. Assume that you are purchasing for Scarlet Oaks Vocational School, 3254 East Kemper Road, Cincinnati, OH. Use today's date and show the date of delivery as one week from that date.

1. Purchase Order No. 1493

   To: Hands Packing Co.
   8567 Spring Grove Ave.
   Cincinnati, OH

   Description–Ground Beef–Chuck
   85% lean 15% suet
   Medium Grind

   Quantity–165 pounds   Unit price $.99 per pound

   _____

   Description–10-ounce Sirloin Steaks–Choice
   1½-inch tail
   Frozen
   Quantity–260 pounds   Unit price $2.45 per pound

   _____

   Description–14 pounds Pork Loins–Spine bones removed; rib and loin end are separated leaving 2 ribs on loin end. Tenderloin left on loin end.

   Quantity–84 pounds   Unit Price $1.26 per pound

   _____

2.   Purchase Order No. 1494

   To: Ideal Bakers Supply
         458 Ross Ave.
         Cincinnati, OH

   Description—Cake flour, 100 pounds

   Quantity 9   Unit Price $15.80 per 100 pounds

   Description—Pastry flour, 100 pounds

   Quantity 9   Unit Price $13.70 per 100 pounds

   Description—Powdered sugar, 10X, 25-pound bag

   Quantity 5   Unit Price $3.95 per bag

   Description—Meringue powder, 10-pound box

   Quantity 3   Unit Price $5.48 per box

---

3.   Purchase Order No. 1495

   To: Deluxe Foods
         6870 High St.
         Hamilton, OH

   Description—Tuna fish, light meat, chunk 24–13-ounce cans

   Quantity 12 cases   Unit Price $14.70 per case

   Description—Coffee, drip grind, 12–2-pound cans

   Quantity 5 cases   Unit Price $48.00 per case

   Description—Sliced pineapple, 50 count, 6–#10 cans

   Quantity 13 cans   Unit Price $12.90 per case

   Description—Pear halves, 50 count, 6–#10 cans

   Quantity 7 cases   Unit Price $10.50 per case

---

4. Purchase Order No. 1496

    To: Surk's Meat Packing Co.
    1520 Eastern Ave.
    Covington, KY

    Description–Boston Butt, average, 4 pounds, cottage butt and blade
    bone removed.

    Quantity 48 pounds   Unit Price $1.18 per pound

    Description–Ham, average, 12 pounds, shank bone removed

    Quantity 36 pounds   Unit Price $1.34 per pound

    Description–Mett Sausage, 6 to 1 pound

    Quantity 26 pounds   Unit Price $1.12 per pound

    Description–Bacon, lean, 28 slices to 1 pound

    Quantity 28 pounds   Unit Price $.89 per pound

    ---

5. Purchase Order No. 1497

    To: Norwood Food Co.
    149 Dexter Ave.
    Norwood, OH

    Description–Whole tomatoes, solid pack, 6–#10 cans

    Quantity 5 cases   Unit Price $7.26 per case

    Description–Powdered sugar, 6X, 24–1-pound boxes

    Quantity 12 cases   Unit Price $2.94 per case

    Description–Asparagus spears, all white 24–1-pound cans

    Quantity 7 cases   Unit Price $31.92 per case

    Description–Whole beets, baby (rose bud), 120 count, 6–#10 cans

    Quantity 4   Unit Price $15.00 per case

UNIT 9
Production Formulas

There are several basic formulas that food service production workers can use to increase their productivity and become more valuable employees. The following 22 formulas are the ones that have been found to be the most helpful. The student should memorize them and learn how they are applied by the use of very simple mathematics.

1. **A pint of liquid is a pound the world around.** This formula is applied when the liquid amount is expressed in weights. It eliminates the time-consuming task of actually weighing the item. For example, if the recipe requires 2 pounds of milk, fill a quart container with milk to get the required 2 pounds. If the recipe requires 4 pounds of water, fill a half-gallon container to get the required 4 pounds. This formula also applies to whole eggs, egg whites, and egg yolks.

2. **Four ounces of powdered milk (dry milk solids) added to 1 quart of water produces 1 quart of liquid milk.** Powdered milk (dry milk solids) is cheaper to buy than whole milk. Therefore, it is often used as a substitute for whole milk in food service production. For example, a vanilla pie filling recipe may require 6 quarts of milk. The chef may instruct that 4 quarts of fresh whole milk be used and 2 quarts of reconstituted dry milk. In this case, 8 ounces of dry milk are dissolved in 2 quarts of water. The quality of the product remains the same, but the food cost is reduced. When purchasing dry milk (dry milk solids), purchase the kind that has a high degree of solubility (ability to dissolve) so it blends quickly when mixed with water.

3. **To prepare flavored gelatin, use 1 cup of gelatin powder to each quart of liquid (water or fruit juice).** This formula provides a gelatin that is stiff enough to keep its shape if unmolded properly. It also provides excellent eating qualities. If the gelatin mold is to be displayed in a slightly warm room, a little more gelatin should be added. When preparing the gelatin, half the liquid should be hot and the other half ice cold. Thoroughly dissolve the gelatin powder in the hot liquid before adding the cold liquid.

4. **To prepare aspic (clear meat, fish or poultry jelly) or chaud-froid (jellied white,**

meat or poultry sauce) use 6 ounces of unflavored gelatin to each gallon of liquid. This produces an aspic or chaud-froid that adheres tightly to the product it covers. To determine the amount of gelatin needed, use 6 times the number of gallons required. For example, if 3 gallons of aspic are needed, $6 \times 3 = 18$ ounces (1 pound 2 ounces) of gelatin are required.

5. **To prepare baked rice, use a ratio (proportion) of 2 parts liquid to every 1 part raw rice by volume.** If the rice is baked properly (approximately 20 minutes in a 400°F oven), it produces an excellent finished product. To determine the amount of liquid needed when preparing a specific amount of rice, multiply the amount of rice by two. For example, if 2 quarts of rice are to be baked, 4 quarts (1 gallon) of liquid would be needed ($2 \times 2 = 4$).

6. **To boil barley, use a ratio (proportion) of 4 parts liquid to every 1 part raw barley by volume.** For example, if it is desired to boil 2 quarts of barley, 8 quarts (2 gallons) of liquid are required ($2 \times 4 = 8$).

7. **To prepare chicken, beef, or ham stock using the flavored soup bases, use ½ cup (or 4 ounces) of the base for every gallon of water.** For example, if 3 gallons of stock are needed, 12 ounces of base ($4 \times 3 = 12$) or 1½ cups ($½ \times 3 = 1½$) are dissolved into 3 gallons of water.

8. **To prepare a meringue using meringue powder, whip together 5 ounces of meringue powder, 1 quart of water, and 3 pounds of granulated sugar.** This mixture produces approximately 3 gallons of meringue.

9. **To prepare pudding using the pudding powder mix, use 3¼ ounces of powder for every pint of milk.** Fox example, if 2 quarts of pudding are required, 3¼ is multiplied by 4 (4 pints are contained in 2 quarts).

$$3¼ \times \frac{4}{1} = \frac{13}{4} \times \frac{4}{1} = 13$$

10. **To thicken fruit pie filling, use 4 to 5 ounces of starch for every quart of fruit juice.**

11. **To convert dry nondairy creamer to liquid, mix 1 pint of dry creamer to 1 quart of hot water.**

12. **To cook pasta (spaghetti, mostaccioli, macaroni, etc.), use 1 gallon of boiling water for every 1 pound of pasta.** For example, if cooking 3½ pounds of spaghetti, 3½ gallons of boiling water are required. This formula can also be used for egg noodles.

13. **To every 1 pound of coffee, use 2½ gallons of water.** For example, 2½ gallons of liquid coffee equals approximately 60 to 65 cups of coffee.

14. **To prepare mashed potatoes using the dry instant potato powder, use 1 pound 13 ounces of the powder for each gallon of water, milk, or water-milk mixture.** For example, if 4 gallons of mashed potatoes are to be prepared, multiply $4 \times$ 1 pound 13 ounces to determine the amount

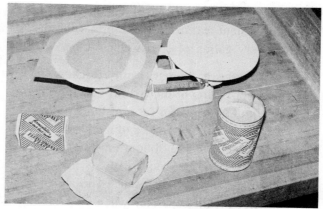

Fig. 9-1 This photo shows the amount of dry yeast needed to replace one pound of compressed yeast.

of instant potato powder needed. First, the pounds must be converted into ounces before multiplication can take place. 1 pound 13 ounces = 29 ounces. 4 × 29 = 116 ounces. The 116 ounces needed must then be divided by 16 to determine pounds and ounces. 116 ÷ 16 = 7¼ pounds or 7 pounds 4 ounces.

15. **If dry yeast, figure 9-1, must be substituted for compressed yeast, 40 percent of the compressed yeast amount is dry yeast. The remaining 60 percent is made up of water.** For example, a recipe may require the use of 1 pound of compressed

yeast (16 ounces). 40 percent of 16 = 6.4 ounces. Use 6¼ ounces of dry yeast. The remaining 60 percent is water. 60 percent of 16 = 9.6 ounces. Use 9¼ or 9½ ounces of water, since the baker's scale is graduated in ¼ ounces. This very slight difference in weight has no effect on the recipe.

16. **To prepare iced tea, figure 9-2, using large 1-ounce tea bags, use the 1-2-3 method.** That is, steep 2 ounces of tea (two bags) in 1 quart of scalding hot water. Steep for approximately 6 minutes. Remove

Fig. 9-2 Student preparing iced tea using the formula given in this unit.

bags and add 3 quarts of cold water. If using instant tea, use 1 ounce of instant tea for every gallon of cold water.

17. **To prepare a proper roux (thickening agent consisting of melted fat and flour), use equal amounts of fat and flour.** For example, if using 8 ounces of melted shortening, then 8 ounces of flour must be added to prepare a proper roux. Roux is used to thicken most soups and sauces because it does not break (separate) when excessive heat is applied to the product.

18. **When cooking dried legumes (dried vegetables), use a ratio of four to one, or one gallon of water to every one quart of legumes.** For best results, soak the legumes overnight before cooking. Cook by simmering.

19. **To prepare pan grease, thoroughly mix together 8 ounces of flour to every 1 pound of shortening.** Pan grease is used to prepare pans for certain baked goods which require that pans be greased and dusted with flour.

20. **To prepare a fairly rich egg wash, mix together 4 whole eggs to every quart of milk.** Egg wash is used quite often in the commercial kitchen. Its most popular use is in the breading procedure. The procedure for breading is to pass an item through seasoned flour, egg wash, and then bread crumbs. Most fried items are breaded.

21. **The suggested amount of liquid to use in the preparation of pie dough is 1 quart of liquid to every 4 pounds of flour.** For example, if 6 pounds of flour is used in a formula, approximately 1½ quarts of liquid should be used.

$$
\begin{array}{r}
1.5 = 1\tfrac{1}{2}\ \text{quarts} \\
4\overline{)6.0} \\
\underline{4} \\
2\ 0 \\
\underline{2\ 0}
\end{array}
$$

22. **To prepare a pie filling using fresh fruit, the amount of liquid used is based on the amount of fresh fruit contained in the formula.** Usually 65% of the fresh fruit amount will provide a sufficient amount of liquid. For example, if 10 pounds of fresh fruit is being used, convert the 10 pounds to 160 ounces ($16 \times 10 = 160$). Take 65% of 160 ounces to determine the amount of liquid to use.

$$
\begin{array}{r}
160 \\
\times\ .65 \\
\hline
800 \\
\underline{960} \\
104.00
\end{array}
$$
ounces or 6 pounds 8 ounces of liquid

ACHIEVEMENT REVIEW
PRODUCTION FORMULAS

1. Determine the amount of liquid required if a pie dough formula contained the following amounts of flour.

   a. 10 pounds
   b. 12 pounds
   c. 16 pounds
   d. 18 pounds

2. Determine the amount of liquid required if the following amounts of fresh fruit are used.

   a. 5 pounds fresh peaches
   b. 7 pounds 8 ounces fresh apples
   c. 8 pounds 12 ounces fresh apricots
   d. 9 pounds 6 ounces fresh apples

3. Determine the amount of dry milk required to produce the following amounts of liquid milk.

   a. 1½ gallons liquid milk
   b. 3 gallons liquid milk
   c. 3½ quarts liquid milk
   d. 4¾ gallons liquid milk

4. Determine the amount of flavored gelatin powder needed to prepare the following amounts of flavored gelatin.

   a. 2 gallons flavored gelatin
   b. 2½ quarts flavored gelatin
   c. 3 gallons flavored gelatin
   d. 3 pints flavored gelatin

5. Determine the amount of unflavored gelatin required to jell the following amounts of aspic.

   a. 6 quarts aspic
   b. 1¾ gallons aspic
   c. 3½ gallons aspic
   d. 2½ gallons aspic

6. Determine the amount of liquid needed to prepare the following amounts of raw rice.

   a. 1 pint raw rice
   b. 3 pints raw rice
   c. 1½ gallons raw rice
   d. 1½ quarts raw rice

7. Determine the amount of liquid needed to prepare the following amounts of raw barley.

   a. 1 pint raw barley
   b. 3 pints raw barley
   c. 3 quarts raw barley
   d. 1 cup raw barley

8. Determine the amount of base needed to prepare the following amounts of stock.

   a. 2 gallons stock
   b. 2¾ gallons stock
   c. ½ gallon stock
   d. 8¾ gallons stock

9. Determine the amount of pudding powder needed to prepare the following amounts of pudding.

   a. 3 quarts pudding
   b. 2½ quarts pudding
   c. 6 quarts pudding
   d. 1½ quarts pudding

10. Determine the amount of starch required to thicken the following amounts of fruit juice.

    a. 3 quarts fruit juice
    b. 1½ quarts fruit juice
    c. 2 gallons fruit juice
    d. 1½ gallons fruit juice

11. Determine the amount of dry, nondairy creamer required to produce the following amounts of liquid cream.

    a. 2 gallons liquid cream
    b. 1½ gallons liquid cream
    c. ½ gallon liquid cream
    d. 3½ gallons liquid cream

12. How many gallons of liquid coffee can be produced from the following amounts of coffee?

    a. 3 pounds coffee
    b. 6 pounds coffee
    c. 4 pounds coffee
    d. 5 pounds coffee

13. Determine the amount of dry instant potato powder needed to prepare the following amounts of mashed potatoes.

    a. 2 gallons mashed potatoes
    b. 6 gallons mashed potatoes
    c. 2½ gallons mashed potatoes
    d. 4½ gallons mashed potatoes

14. For the following amounts of compressed yeast, determine the proper amount of dry yeast and water to use if substituting dry yeast for compressed yeast.

    a. 12 ounces compressed yeast
    b. 10 ounces compressed yeast
    c. 6 ounces compressed yeast
    d. 1 pound 4 ounces compressed yeast

15. How many pounds are contained in the following amounts of liquid?

    a. 1½ gallons
    b. 3 gallons
    c. 2¾ gallons
    d. 4¼ gallons

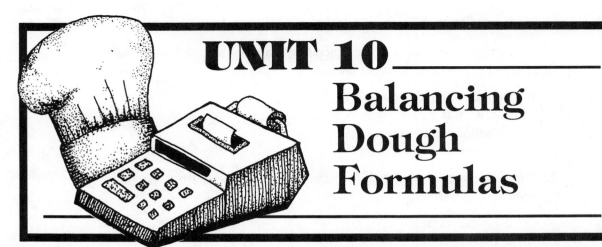

UNIT 10
Balancing Dough Formulas

The ingredients in baking formulas must be balanced if the finished product is to possess all the qualities necessary to please the customer and to warrant return sales. Most formulas in use today have been developed in research laboratories operated by the companies that manufacture the products used in bakeshops. The formulas are used to test their products, which are then distributed to bakers in the hope that they will use them.

The formulas are designed to balance all of the ingredients with flour, the main ingredient. Each minor ingredient is a percentage of the main ingredient. Industry standards determine the percentage of each ingredient in most of the popular formulas to ensure that the formula is balanced. For example, let us take pie dough.

A good pie must have a tender, flaky pie crust. Tenderness and flakiness depend on the ratio of shortening to flour (the ratio between two quantities is the number of times one contains the other). The higher the percentage of shortening to flour by weight, the more tender and fragile is the baked crust. So the percentage of shortening used is determined by how the dough will be handled when the pies are made and by whether the pie will be shipped from one location to another. This is explained so you can understand the variation in shortening percentage for a balanced pie dough formula.

The suggested percentages used for a balanced pie dough formula are as follows:

Pastry Flour	100%
Shortening	65 to 75%
Salt	2 to 3%
Sugar	1 to 3%
Dry Milk	1 to 2%
Water	25 to 35%

The flour represents 100% because the amounts of the other ingredients are based on the amount of flour used. To check the percentages of each ingredient in a pie dough formula to see if it is balanced and will produce a quality crust, divide the weight of the flour into the weight of each ingredient. For example, let us take the following formula and see if it will balance with the percentages suggested.

Pie dough ingredients to yield approximately 20–8-inch pies

10 pounds   Pastry Flour

7 pounds	Shortening
4 ounces	Salt
2 ½ ounces	Dry Milk
2 ½ ounces	Sugar
3 pounds 6 ounces	Water

Convert 10 pounds of flour to 160 ounces:

$$10 \times 16 = 160$$

Convert 7 pounds of shortening to 112 ounces:

$$7 \times 16 = 112$$

To find a percent, divide the whole into the part. In this case 160 ounces of flour represents the whole; the 112 ounces of shortening represents the part.

$$160\overline{)112.000} = 70\% \text{ shortening}$$
$$\underline{112\ 0}$$

To find the remaining percents, continue to divide the whole (160 ounces) into the amount of each ingredient. The results are as follows:

Pastry Flour	100%
Shortening	70%
Salt	2.5%
Dry Milk	1.5%
Sugar	1.5%
Water	33.7%

The figures show that this formula will balance with the suggested percentages and will produce an excellent pie dough.

Let us take one more example to make sure we understand the concept of balancing a formula

and the importance of working with formulas that balance.

The suggested percentages used for a balanced, medium-rich, soft dinner roll formula is as follows:

Bread Flour	100%
Sugar	10 to 16%
Shortening	15 to 20%
Salt	1 to 2%
Dry Milk	5 to 8%
Whole Eggs	5 to 8%
Water	50 to 60%
Compressed Yeast	7 to 10%

The soft dinner roll dough to be checked is as follows:

Soft dinner roll dough ingredients to yield approximately 15 dozen rolls

1 pound	Granulated Sugar
1 pound 4 ounces	Shortening
8 ounces	Dry Milk
2 ounces	Salt
6 ounces	Whole Eggs
4 pounds	Cold Water
10 ounces	Compressed Yeast
7 pounds	Bread Flour

Convert 7 pounds of flour to 112 ounces ($7 \times 16 = 112$ ounces).

Convert 1 pound of sugar to 16 ounces ($16 \times 1 = 16$ ounces).

To find a percent, divide the whole into the part. In this case 112 ounces of flour represents the whole, and the 16 ounces of sugar represents the part.

$$
\begin{array}{r}
.142 = 14.2\%\ \text{sugar} \\
112\overline{)16.000} \\
\underline{112} \\
480 \\
\underline{448} \\
320 \\
\underline{224} \\
96
\end{array}
$$

To find the remaining percents, continue to divide the whole (112 ounces) into the amount of each ingredient. The results are as follows:

Granulated Sugar	14.2%
Shortening	17.8%
Dry Milk	7.1%
Salt	1.7%
Whole Eggs	5.3%
Cold Water	57.1%
Compressed Yeast	8.9%
Bread Flour	100%

The figures show that this formula will balance with the suggested percentages and will produce a quality, medium-rich, soft dinner roll dough. Keep in mind that the suggested percentages for yeast doughs will vary because there are two types: lean dough and rich-sweet dough.

ACHIEVEMENT REVIEW
BALANCING FORMULAS

Using the suggested ingredient percentages for a balanced pie dough or roll dough formula given on pages 120 and 121, determine if the following formulas balance.

1. Mealy-Type Pie Dough Ingredients

10 pounds		Pastry Flour	_____	100%
7 pounds	8 ounces	Shortening	_____	_____
	5 ounces	Salt	_____	_____
	3 ounces	Sugar	_____	_____
2 pounds	8 ounces	Cold Water	_____	_____
	3 ounces	Dry Milk	_____	_____

2. Short-Flake-Type Pie Dough Ingredients

10 pounds		Pastry Flour	_____	__100%__
7 pounds	8 ounces	Shortening	_____	_____
	5½ ounces	Salt	_____	_____
3 pounds		Cold Water	_____	_____
	6 ounces	Sugar or Corn Sugar	_____	_____
	3 ounces	Dry Milk	_____	_____

3. Long-Flake-Type Pie Dough Ingredients

12 pounds		Pastry Flour	_____	__100%__
7 pounds	12 ounces	Shortening	_____	_____
	4 ounces	Salt	_____	_____
5 pounds	4 ounces	Cold Water	_____	_____
	4 ounces	Sugar	_____	_____
	3 ounces	Dry Milk	_____	_____

4. Soft Dinner Roll Dough Ingredients

1 pound	4 ounces	Granulated Sugar	_____	_____
1 pound	6 ounces	Shortening	_____	_____
	2 ounces	Salt	_____	_____
	6 ounces	Dry Milk	_____	_____
	8 ounces	Whole Eggs	_____	_____

7 pounds	10 ounces	Bread Flour	_____	__100%__
4 pounds		Water	_____	_____
	10 ounces	Compressed Yeast	_____	_____

5. Golden Dinner Roll Dough Ingredients

1 pound		Granulated Sugar	_____	_____
	12 ounces	Whole Eggs	_____	_____
	5 ounces	Salt	_____	_____
	8 ounces	Dry Milk	_____	_____
	10 ounces	Compressed Yeast	_____	_____
5 pounds		Cold Water	_____	_____
9 pounds		Bread Flour	_____	__100%__
1 pound		Pastry Flour	_____	_____
1 pound		Butter Flavored Shortening	_____	_____

UNIT 11
Finding Approximate Recipe Yields

Yield means to produce. When used in reference to a recipe or formula, it provides the preparer with an approximate guide to the number of portions, servings, or units a particular recipe or formula will produce. The yield is an important feature of any recipe or formula because it is a tool for controlling food production and food cost.

The yield for some recipes or formulas is found by preparing a certain amount, determining the serving portion, and measuring it to see what it will produce. The yield for other recipes, such as cake or muffin batters, roll or sweet doughs, pie fillings, and some cookie doughs, can be determined by taking the total weight of all ingredients used in the preparation and dividing that figure by the weight of an individual portion or unit. Let us take the formulas of a white cake and of a roll dough to show how an approximate yield can be obtained by this method.

White Cake Ingredients

2 pounds 8 ounces	Cake Flour
1 pound 12 ounces	Shortening
3 pounds 2 ounces	Granulated Sugar
1½ ounces	Salt
2½ ounces	Baking Powder

14 ounces	Water
2½ ounces	Dry Milk
10 ounces	Whole Eggs
1 pound	Egg Whites
1 pound	Water
	Vanilla to Taste

The total weight of all ingredients is 11 pounds 4½ ounces. Each cake is to contain 14 ounces of batter. The first step is to convert the weight of all ingredients to ounces, since only like things can be divided: 11 pounds 4½ ounces contains 180½ ounces. The second step is to divide the weight of one cake (14 ounces) into the total weight of all ingredients.

$$
\begin{array}{r}
12. \\
14\overline{)180.5} \\
\underline{14} \\
40 \\
\underline{28} \\
12
\end{array}
$$

There is no reason to carry the division any further because only figures on the left side of the decimal point are whole numbers. So 12 cakes, each containing 14 ounces of batter, were realized from this recipe.

Soft Dinner Roll Dough Ingredients

1 pound 4 ounces	Granulated Sugar
1 pound 4 ounces	Shortening
2 ounces	Salt
6 ounces	Dry Milk
6 ounces	Whole Eggs
7 pounds 8 ounces	Bread Flour
4 pounds	Water
10 ounces	Compressed Yeast

The total weight of all ingredients is 15 pounds 8 ounces. Each roll is to weigh 1½ ounces. The first step is to convert the weight of all ingredients to ounces, since only like things can be divided: 15 pounds 8 ounces contains 248 ounces. Now, dividing the total weight by the weight of one roll gives

$$
\begin{array}{r}
165. \\
1.5\overline{)248.0} \\
\underline{15} \\
98 \\
\underline{90} \\
80 \\
\underline{75} \\
5
\end{array}
$$

Thus the yield is 165 rolls, or 13¾ dozen rolls.

ACHIEVEMENT REVIEW
FINDING AN APPROXIMATE YIELD

1. Determine the approximate yield of the following formula if each roll is to contain 1½ ounces of dough.

   Soft Dinner Roll Dough Ingredients

1 pound		Granulated Sugar
1 pound	4 ounces	Shortening
	8 ounces	Dry Milk
	2 ounces	Salt
	6 ounces	Whole Eggs
	6 ounces	Compressed Yeast
4 pounds		Cold Water
7 pounds		Bread Flour

2. Determine the approximate yield of the following formula if each coffee cake is to contain a 12-ounce unit of sweet dough.

   Coffee Cake Ingredients

1 pound		Granulated Sugar
1 pound		Golden Shortening
	1 ounce	Salt
3 pounds		Bread Flour

1 pound	8 ounces	Pastry Flour
	12 ounces	Whole Eggs
	4 ounces	Dry Milk
2 pounds		Water
8 ounces		Compressed Yeast
		Mace to Taste
		Vanilla to Taste

3. Determine the approximate yield of the following formula if each Danish roll is to contain 2 ounces of Danish pastry dough.

Danish Pastry Dough Ingredients

	12 ounces	Granulated Sugar
	12 ounces	Golden Shortening
	1¾ ounces	Salt
3 pounds		Bread Flour
1 pound	8 ounces	Pastry Flour
1 pound		Whole Eggs
	4 ounces	Dry Milk
2 pounds		Water
	8 ounces	Yeast
2 pounds	8 ounces	Golden Shortening
		(roll in)

4. Determine the approximate yield of the following formula if each cookie is to contain 1½ ounces of dough.

Fruit Tea Cookies

1 pound	6 ounces	Shortening
1 pound	6 ounces	Powdered Sugar
2 pounds	8 ounces	Pastry Flour
	2 ounces	Liquid Milk
	6 ounces	Raisins, Chopped
	2 ounces	Pecans, Chopped
	3 ounces	Pineapple, Chopped
	2 ounces	Peaches, Chopped
	8 ounces	Whole Eggs
	¼ ounce	Baking Soda
	¼ ounce	Vanilla
	½ ounce	Salt

5. Determine the approximate yield of the following formula if each cookie is to contain ¾ ounces of dough.

Danish Butter Cookies Ingredients

1 pound	8 ounces	Granulated Sugar
	12 ounces	Shortening
	12 ounces	Butter
	¼ ounce	Salt
2 pounds	6 ounces	Cake Flour
	¼ ounce	Baking Powder
	½ ounce	Dry Milk
	5 ounces	Whole Eggs
	4 ounces	Water
		Vanilla to Taste

6. Determine the approximate yield of the following formula if each cake is to contain 12 ounces of batter.

Semi-Sponge Cake Ingredients

2 pounds	8 ounces	Cake Flour
3 pounds		Granulated Sugar
	1½ ounces	Salt
	2½ ounces	Baking Powder
	¼ ounce	Baking Soda
1 pound		Egg Yolks
1 pound		Whole Eggs
	8 ounces	Water
	4 ounces	Dry Milk
	8 ounces	Salad Oil
1 pound	4 ounces	Water
		Vanilla to Taste

7. Determine the approximate yield of the following formula if each cake is to contain 13 ounces of batter.

Devil's Food Cake Ingredients

2 pounds	8 ounces	Cake Flour
1 pound	6 ounces	Shortening

3 pounds	8 ounces	Granulated Sugar
	8 ounces	Cocoa
	1½ ounces	Salt
	¾ ounces	Baking Soda
	1½ ounces	Baking Powder
	6 ounces	Dry Milk
1 pound	4 ounces	Water
1 pound	14 ounces	Whole Eggs
1 pound	9 ounces	Water
	¼ ounce	Vanilla

8. Determine the approximate yield of the following formula if each cake is to contain 11 ounces of batter.

Yellow Cake Ingredients

2 pounds	8 ounces	Cake Flour
1 pound	6 ounces	Shortening
3 pounds	2 ounces	Granulated Sugar
	1 ounce	Salt
	1¾ ounces	Baking Powder
	4 ounces	Dry Milk
1 pound	4 ounces	Water
1 pound	10 ounces	Whole Eggs
	12 ounces	Water
	¼ ounce	Vanilla

9. Determine the approximate yield of the following formula if each pie is to contain 1 pound 14 ounces of filling.

Cranberry-Apple Pie Filling Ingredients

7 pounds		Apples, Canned
2 pounds		Water or Apple Juice
1 pound	8 ounces	Granulated Sugar
	¼ ounce	Cinnamon
	⅛ ounce	Nutmeg
	¼ ounce	Salt
	4 ounces	Starch
	6 ounces	Water

		Whole Cranberry
2 pounds		Sauce
	4 ounces	Corn Syrup
	¼ ounce	Lemon Juice

10. Determine the approximate yield of the following formula if each pie is to contain 1 pound 10 ounces of filling.

Vanilla Pie Filling Ingredients

12 pounds		Liquid Milk
4 pounds		Granulated Sugar
1 pound		Cornstarch
	¼ ounce	Salt
2 pounds		Whole Eggs
	6 ounces	Butter
	¼ ounce	Vanilla

UNIT 12
Finding Meat Cut Percentages

It used to be a common practice for the owner and the chef of a food service operation to visit the local slaughter house once a week to select the sides or saddles of beef, veal, pork, or lamb, they wished to purchase for their operation. Today, purchasing animals by the side or saddle is pretty much limited to large operations—those with commissaries that process food to be shipped to their various units.

A side is half of the complete carcass. A saddle, used in reference to the lamb carcass, is the front or hind half of the complete carcass that is cut between the twelfth and thirteenth ribs of the lamb. The side or saddle is blocked out into wholesale or primal cuts at the food service

operation. The food service operator must know what percentage of the side or saddle goes into each of the wholesale or primal cuts so that he or she can determine whether it is advantageous to purchase meat in this form.

Here is an example of how a side of beef is blocked out and how the percentage of each wholesale or primal cut is found. For the side of beef in figure 12-1, find the percentage of each wholesale or primal cut.

To find a percentage, divide the whole into the part. In this case the whole is represented by the total weight of the side of beef, 222 pounds. The part is represented by the weight of each individual wholesale or primal cut. For

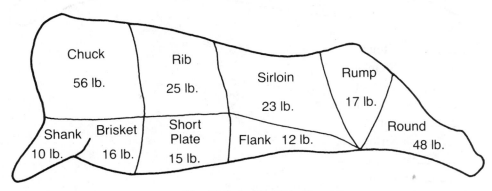

Fig. 12-1 A side of beef

example, the percentage of chuck is found as follows:

$$.252 = 25.2\%$$
$$222\overline{)56.000}$$
$$\underline{444}$$
$$1160$$
$$\underline{1110}$$
$$500$$
$$\underline{444}$$
$$56$$

The percentages of the other beef cuts are found by following the same procedure. Carry three places behind the decimal.

25.2% Chuck	10.3% Sirloin
4.5% Shank	5.4% Flank
7.2% Brisket	7.6% Rump
6.7% Short Plate	21.6% Round
11.2% Rib	

ACHIEVEMENT REVIEW
FINDING THE PERCENTAGE OF BEEF, PORK, VEAL AND LAMB WHOLESALE OR PRIMAL CUTS

1. Find the percentage of each wholesale or primal cut of beef that makes up the side of beef shown in the figure.

Total weight of the side _____

_____% Chuck
_____% Shank
_____% Brisket
_____% Short Plate
_____% Rib
_____% Sirloin
_____% Flank
_____% Rump
_____% Round

2. Find the percentage of each wholesale or primal cut of veal that makes up the side of veal shown in the figure.

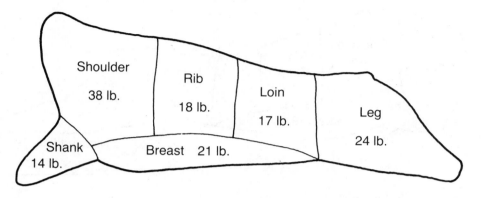

Total weight of side _____

_____% Shoulder   _____% Rib
_____% Shank   _____% Leg
_____% Breast   _____% Loin

3. Find the percentage of each wholesale or primal cut of pork that makes up the side of pork shown in the figure.

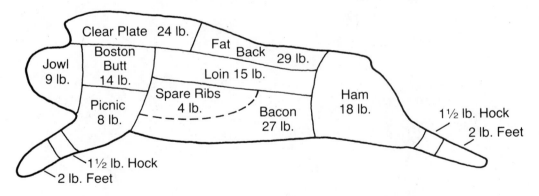

Total weight of side _____

_____% Jowl   _____% Feet
_____% Boston Butt   _____% Loin
_____% Clear Plate   _____% Bacon
_____% Ham   _____% Spare Ribs
_____% Picnic   _____% Fat Back
_____% Hock

4. Find the percentage of each cut of lamb that makes up the foresaddle and hindsaddle of lamb shown in the figure.

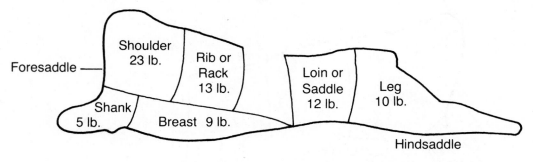

Total Weight of Side _____

_____% Shoulder

_____% Shank

_____% Breast

_____% Rib or Rack

_____% Loin or Saddle

_____% Leg

_____% Foresaddle

_____% Hindsaddle

PART THREE

Service Crew—
Operational
Procedures

The service crew consists of the employees that work in the front of the house (the dining room). They include the hostess or captain, maitre d' hotel (referred to as maitre d'), waiter, waitress, cashier, food checker or expediter, and busboy. Each of these employees is involved with some form of mathematics.

The *hostess* or *captain* greets the guests as they enter the dining room, shows them to their seats, and hands them a menu. At this point, the waiter or waitress takes over. The hostess or captain is involved in mathematics in the areas of organizing work schedules for the waiter or waitress, assigning service stations, keeping the reservation book up to date, and controlling seating arrangements.

The *maitre d'* is in charge of the dining room service. This member of the crew is involved in every phase of the operation, and therefore must possess knowledge of all of the mathematics involved in dining room operation.

The *waiters* or *waitresses* are the ones who serve the meal. They must keep track of all food and beverages purchased and enter prices on the guest check. Besides the cashier, they are more involved with mathematics than any other member of the service crew.

The *cashier* controls the cash. This job includes receiving payment of the sales checks, making change, and filling out the cashier's daily worksheet.

The *food checker* or *expediter* is responsible for all foods that leave the kitchen. This includes checking all the trays that leave the kitchen area and making sure that only the foods ordered are on the tray. The food checker also keeps track of the amount of each item served. In this way, it is possible to check which items are selling, and double check the sales checks received by the cashier.

The *busboy* removes all used dishes from the table, resets the table, and takes the used dishes from the dining room to the dishwashing area. He also assists the waiter or waitress by carrying trays of food from the kitchen to the dining room. The only mathematics involved in this job is counting the number of pieces of silverware, china, and glassware needed to set the tables.

The mathematics involved in dining room operation consists of the following operational procedures. The service crew should be familiar with each of these items.

- Sales Checks
- Cashier's Daily Worksheet
- Tipping
- Using The Calculator

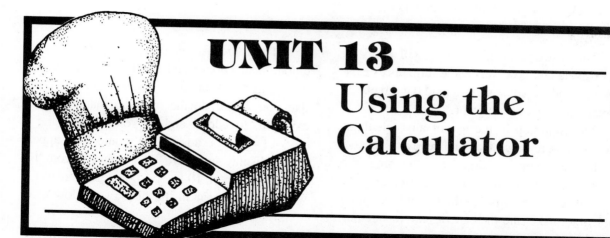

UNIT 13

Using the Calculator

Although the electronic calculator has been widely used for many years, many teachers did not want their students to use them because they thought the calculators would hinder the students' learning of proper math skills. Students would become lazy if only the fingers were exercised instead of the brain. Consequently, the electronic calculator was forbidden in many classrooms. However, teachers now encourage the use of the calculator. Because calculators are so popular in business, schools, and homes, companies are producing them in large quantities. Thus, with production up, unit cost is reduced, profits are constant, and the selling price is lowered. The cost of a calculator is now a fraction of what it once was.

The face of the small hand-held calculator shown in figure 13-1 contains the following keys:

ON	Press to turn calculator on.
C	When power is on, this key is pressed to clear the calculator.
CE	Press to clear an incorrect keyboard entry.
0 to 9	Numeral entry keys
.	Decimal point key
=	Equals or result key
+	Plus or addition key

−	Minus or subtraction key
×	Multiplication key
÷	Division key
%	Percentage key
M+	Memory plus key
M−	Memory minus key
RM	Memory recall
CM	Memory clear key

Fig. 13-1 An example of a small hand calculator (Courtesy of Sharp Electronics Corporation, EL-233)

The instructions given here will apply to most calculators; however, some differences exist between models, so be sure to read the directions for your particular calculator.

The first step in the use of any calculator is to make sure it is clear and ready to receive calculations. This is done by touching the ON/C key, meaning "on" and "clear." At this point it is also wise to do a simple addition problem to make sure the batteries are functioning properly.

The four basic methods of operation—addition, subtraction, multiplication, and division—are carried out on the calculator in the order that you would do the problem manually. For example, to add 6 and 4, you would enter 6 + 4 = . The answer 10 would then appear in the display window. To subtract 4 from 10, you would enter 10 − 4 = . The answer 6 would then appear in the display window.

Try the following practice exercises to see if you have mastered how to use a calculator for the four basic operations. Correct answers are given.

1. Addition practice exercise
   a. 37 + 46 + 54 =   *Ans.* 137
   b. 48 + 52 + 78 =   *Ans.* 178
   c. 3463 + 225 + 2218 + 4560 =   *Ans.* 10466
   d. 32.5 + 519.43 + 2226.06 + 18.03 =   *Ans.* 796.02
   e. 26423 + 22.08 + 2946 + 3220 + 445.046 =   *Ans.* 33056.126

2. Subtraction practice exercise
   a. 33682 − 18620 =   *Ans.* 15062
   b. 3895.28 − 1620.29 =   *Ans.* 2274.99
   c. 48920.56 − 32826.69 =   *Ans.* 16093.87
   d. $8668.78 − $4878.28 =   *Ans.* $3790.50
   e. 956 − 482.739 =   *Ans.* 473.261

3. Multiplication practice exercise
   a. 86 × 256 =   *Ans.* 22016
   b. 1620 × 62 × 18 =   *Ans.* 1807920
   c. 4482 × 22 × 6.8 =   *Ans.* 670507.2
   d. 46.5 × 7 × 12.2 × 18.4 =   *Ans.* 73068.24
   e. $438.75 × 34.5 =   *Ans.* $15136.875

4. Division practice exercise
   a. 2175 ÷ 15 =   *Ans.* 145
   b. 7137 ÷ 156 =   *Ans.* 45.75
   c. 6256.25 ÷ 175 =   *Ans.* 35.75

d. 82.9 ÷ .45 =

e. 6.5 ÷ .25 =

Ans. 184.222

Ans. 26

Chain calculations involve a series of numbers and a variety of math procedures. Many chain calculations involve all four basic operations. For example,

$$29 + 120 - 38 \times 3 \div 3 = 111$$

5. Chain calculations practice exercise

a. 97 + 120 − 38 × 5 ÷ 4 =

b. 1440 − 1200 + 45 ÷ 2 × 2 =

c. 395 × 42 − 225 + 448 =

d. $53785.25 − $32726.85 × .05 =

e. 596 × 58 − 24568 + 1420.6 =

Ans. 223.75

Ans. 285

Ans. 16813

Ans. $1052.92

Ans. 11420.6

The constant function may be used to multiply or divide repeatedly by the same number. The constant is entered first when multiplying and becomes the multiplier. The multiplier remains in the calculator as a new multiplicand is entered. The $\boxed{=}$ key is depressed to get the result.

Example: Find the product of the following problems.

What is the amount of sales tax on the following amounts if the 5.5% sales tax rate is used as the constant?

a. $225.00

b. $68.00

c. $2560.90

Remember, 5.5% must be changed to the decimal .055 and entered first, followed by the multiplicand. Pressing the $\boxed{=}$ key will give each product. Do the following examples. Answers are rounded to the nearest cent.

Example A

.055 × $225. = $12.38

Example B

$68. = $3.74

Example C

$2560.90 = $140.85

As you can see, the constant is entered at only one time. New multiplicands are entered without re-entering the constant.

In division, the constant is added after the first dividend is entered. It becomes the divisor and remains in the calculator as each new dividend is entered.

Example: Find the quotient of the following problems. Using 85% as the percent of cost, find the selling price based on the wholesale price of each of the following items:

a. $790.25

b. $48.90

c. $2345.85

Remember that 85% must first be changed to the decimal .85 and entered after the first dividend. Pressing the $\boxed{=}$ key will give each quotient. Do the following examples. Answers are rounded to the nearest cent.

Example A

790.25 ÷ .85 = $929.71

Example B

68 = $80.00

Example C

2560.90 = $3012.82

As you can see, the constant is entered second, and only once, as new dividends are entered.

MULTIPLICATION AND DIVISION BY A PERCENT

Multiplication and division by a percent are functions that are performed similar to the four basic operations in that the functions are performed just as you would express the problem orally. For example, you would say $580 times 5.5%, so on the calculator you enter 580 × 5.5%, and the correct answer appears in the display window as 31.90. Or you can change 5.5% to the decimal .055 and proceed to find the solution by entering 580 × .055 = 31.90. The calculator will automatically place the decimal point, but the dollar sign must be added.

1. Multiplication by a percent exercises
   Try the following practice exercises to see if you have mastered this function. Answers given are rounded off.

   a. $896.25 × 7.5% or
      .075 =   *Ans.* $67.22
   b. $652.40 × 6.7% or
      .067 =   *Ans.* $43.71

   c. $2900.00 × 35% or
      .35 =   *Ans.* $1015.
   d. $7200.00 × 15.6%
      or .156 =   *Ans.* $1123.20
   e. 958.20 × 7.9% or
      .079 =   *Ans.* 75.70

When dividing by a percent, as stated before, the problem is entered just as you would express it orally. For example, you would say $580 divided by 5.5%, and the correct answer appears in the display window as 10545.454; or you can change the 5.5% to the decimal .055 and proceed to find the solution by first entering 580 ÷ .055 = 10545.454. Again, the calculator will automatically place the decimal point, but the dollar sign must be added.

2. Division by a percent exercises
   Try the following practice exercises to see if you have mastered this function. Answers given are rounded off.

   a. $896.25 ÷ 7.5%
      or .075 =   *Ans.* $11950.
   b. $652.40 ÷ 6.7%
      or .067 =   *Ans.* $9737.31
   c. $2900.00 ÷ 35%
      or .35 =   *Ans.* $8285.71
   d. $7200.00 ÷ 15.6%
      or .156 =   *Ans.* $46153.85
   e. 958.20 ÷ 7.9% or
      .079 =   *Ans.* $12129.11

THE MEMORY FUNCTIONS

The memory function is used to retain figures in the calculator. Even when power is turned off, some calculators will retain figures. The memory function makes totaling an invoice or figuring totals simpler when multiplying by a percent. Since the keys of some calculators may vary slightly, it is wise to check the calculator's instruction booklet before using the memory function.

Example: Before starting, press the [C] and [CM] keys to clear the calculator and memory. The following is a step-by-step outline of how to use the memory function to complete an invoice. Each step corresponds to a line of the invoice. The completed invoice is in figure 13-2.

Step 1. 4 [×] 8.95 [M+] ^{M}35.80

Step 2. 3 [×] 15.60 [M+] ^{M}46.80

Step 3. 2 [×] 9.50 [M+] ^{M}19.00

Step 4: 5 [×] 8.40 [M+] ^{M}42.00

Step 5. 2 [×] 18.90 [M+] ^{M}37.80

Step 6. [RM] ^{M}181.40

Step 7. [×] 10 [%] ^{M}18.14

Step 8. [M−] ^{M}18.14 [RM] ^{M}163.26

Step 9. 15 ^{M}15. [M+] ^{M}15.

Step 10. [RM] ^{M}178.26

Distributor:	Haines Food Inc.			Phone 771-8800	

Address: **70 Greenbrier Ave.**
Ft. Mitchell, KY 41017

Distributors of Fine Food Products

Wholesale Only   **Date: April 20, 1987**

No. of Pieces 5		Sales Person Joe Jones		Order No. 2860	Invoice No. J2479
Packed by R.H.		Sold to: Mr. John Doe Street: 120 Elm Ave. City: Covington, KY			
Case	**Pack**	**Size**	**Canned Foods**	**Unit Price**	**Total Amount**
4	6	#10 Can	Sliced Apples	$ 8.95	35.80
3	6	#10 Can	Pitted Cherries	15.60	46.80
2	12	#5 Can	Apple Juice	9.50	19.00
5	24	1 lb.	Cornstarch	8.40	42.00
2	24	#2½ Can	Asparagus	18.90	37.80
			Total Amount		181.40
			Less: Special Discount 10%		18.14
			Total Net Price		163.26
			Plus: Delivery Charge		15.00
			Total Invoice Price		178.26

Fig. 13-2 A completed invoice

Multiplication by a Percent using the Memory Function

Example: Mr. Bills purchased 4 food items at $12.50 each and two food items at $9.60 each. If the sales tax on the total purchase is 6.25%, what was the total cost?

Step-by-step solution

Step 1. 4 $\boxed{\times}$ 12.50 $\boxed{\text{M}+}$ $\boxed{^\text{M}50.00}$

Step 2. 2 $\boxed{\times}$ 9.60 $\boxed{\text{M}+}$ $\boxed{^\text{M}19.20}$

Step 3. $\boxed{\text{RM}}$ $\boxed{^\text{M}69.20}$ $\boxed{\times}$ 6.25 $\boxed{\%}$ $\boxed{^\text{M}4.325}$

Step 4. $\boxed{\text{M}+}$ $\boxed{\text{RM}}$ $\boxed{^\text{M}73.525}$

ACHIEVEMENT REVIEW
ELECTRONIC CALCULATOR FUNCTIONS

Addition

1. 48 + 59 + 212 =

2. 78 + 135 + 389 =

3. 269 + 458 + 678 =

4. 3,263 + 298 + 2,229 + 4,680 =

5. 35.6 + 626.42 + 2430.07 + 15.03 =

Subtraction

6. 23583 − 16420 =

7. 4596.26 − 2725.23 =

8. 56750.35 − 31998.67 =

9. $7567.75 − $5428.46 =

10. 987 − 462.634 =

Multiplication

11. 89 × 368 =

12. 1586 × 82 × 15 =

13. 4556 × 23 × 7.6 =

14. 46.8 × 8 × 13.3 × 16.4 =

15. $656.76 × 32.5 =

Division

16. 1675 ÷ 18 =

17. 6280 ÷ 166 =

18. 8245.25 ÷ 186 =

19. 92.7 ÷ .52 =

20. 7.5 ÷ 23 =

Chain Calculations

21. 67 + 230 − 89 × 6 ÷ 2 =

22. 1260 − 1051 + 290 ÷ 2 × 3 =

23. 365 × 52 − 290 + 545 =

24. $42685.50 − 31628.75 × .05 =

25. 592 × 65 − 28562 + 1325.8 =

The Constant Function

What is the amount of sales tax on the following amounts, if the 6.5% sales tax rate is used as the constant?

26. $286.00

27. $79.50

28. $2460.95

29. $5235.45

30. $8164.34

Using 65% as the percent of cost, find the selling price based on wholesale price of each of the following items.

31. $695.25

32. $58.65

33. $2255.75

34. $3140.35

35. $4480.65

36. Complete the following invoice, using your calculator and its memory function.

Distributor: Millers Food Inc.			Phone 772-9654		
Address: 2033 Elm Ave. Norwood, Ohio 45212					
Distributors of Fine Food Products					
Wholesale Only		Date: April 22, 1987			
No. of Pieces 4		**Sales Person** James Jones		**Order No.** 2861	**Invoice No.** J2480
Packed by R.H.		Sold to: Mr. Tim O'Connell Street: 916 Montague St. City: Cincinnati, Ohio 45202			
Case	**Pack**	**Size**	**Canned Foods**	**Unit Price**	**Total Amount**
5	6	#10 Can	Sliced Pears	$10.95	_____
3	6	#10 Can	Sliced Peaches	12.85	_____
6	12	#5 Can	Tomato Juice	9.50	_____
4	24	1 lb.	Cornstarch	8.40	_____
3	24	#2½ Can	Asparagus	18.90	_____
			Total Amount		_____
			Less: Special Discount 12%		_____
			Total Net Price		_____
			Plus: Delivery Charge		12.00
			Total Invoice Price		_____

Fig. 13-3 Multiplication by a percent using the memory function.

37. Jim Jones purchased 12 items at $12.60 each. Sales tax was 5.5%. What was the total cost?

38. Bob Shirley purchased 8 items costing $6.25 each, and 6 items costing $8.95 each. If the sales tax on the total purchase was 6.5%, what was the total cost?

39. Bill Thompson made purchases costing $10.40, $54.80, $7.35, $8.98, and $.56. If the sales tax was 6.75% of the total amount, what was the total cost?

40. Joe Meyers purchased 15 items at $16.20 each and 4 items at $20.50 each. If the sales tax on the total purchase was 4.25%, what was the total cost?

41. Russ Macke purchased items costing $85.95, $16.75, and $9.56. If the sales tax was 5.75% of the total amount, what was the total cost?

Multiplication by a percent

42. $996.30 × 7.6% =

43. $565.45 × 6.5% =

44. $2800.00 × 15.4% =

45. $7452.85 × 5.8% =

Division by a percent (using ÷ and % keys)

46. $425.60 ÷ 7.5% =

47. $352.75 ÷ 6.3% =

48. $2890.00 ÷ 14.2% =

49. $7252.80 ÷ 5.4% =

50. $956.25 ÷ 7.9% =

UNIT 14
Sales Checks

The sales check (also called the guest check, check, or bill) is the responsibility of the waiter or waitress. Although other employees are guided by the information listed on the check during the dining period, the waiter or waitress is held responsible for it. If a guest walks out without paying, or if any checks are lost, the waiter or waitress may be required to pay the amount due.

At the beginning of the dining period, the waiter or waitress is issued a book of checks with a serial number. Each individual check is numbered consecutively. The waiter or waitress usually signs for each book received. In this way, the management knows if a check is missing and if so, the person who is responsible for it. At the end of the day or at the end of a service period, the checks are put back into the original book form to help detect any errors. Numbering checks and being able to identify the person responsible for each check are also important in checking the daily receipts and assisting the accounting department in finding any errors. Never destroy or discard a check without receiving permission from your supervisor. Remember that controls are necessary in any business operation.

The kind of check used depends upon the type of food service operation and the type of menu. The three types of sales checks generally used are the blank check, partly printed check, and printed check.

BLANK CHECK

When a blank check is used, figure 14-1, the waiter or waitress must be familiar with the items and prices on the menu and must have a neat, legible handwriting. The check must be clear for several reasons: (1) the cook must be able to read it to fill the entree order, (2) the cashier must be able to understand the figures, (3) the manager may wish to examine it (figure 14-2), and (4) the customers want to understand what they are paying for. On a blank sales check, the waiter or waitress is required to record the date, their initials or other identifying mark, table number, and number of persons being served. This information is usually listed at the top of the check. They must write down each item ordered, list prices, give subtotal, add tax, and find the total. It is therefore very important to have an alert, well-informed, experienced service crew when using this type of check.

LaRosa's

Date	Server	Table No.	No. Persons	NO. 487176
1				
2				
3				
4				
5				
6				
7				
8				
9				
10				
11				
12				
13				
14				
15				
16				
17				
18				
19				
20				
21				
22		SUBTOTAL		
23		TAX		
24		TOTAL		
—				NO. 487176

Date	Guest Receipt	Persons	Amount of Check	

LaRosa's

Fig. 14-1 Blank check

PARTLY PRINTED CHECKS

Partly printed checks, figure 14-3, are used in food service operations which have a limited

Fig. 14-2 The manager sometimes examines the sales check before the cashier rings it up.

menu because most of the items listed are best sellers and appear on the menu every day. This type of check speeds service and reduces errors. The standard menu items and the popular beverages that appear each day are listed, but prices are not listed. The waiter or waitress must record all information requested at the top of the check, list quantities and prices, find the subtotal, add taxes, and calculate the total.

PRINTED CHECK

Printed checks, figure 14-4, have the names and prices of all food and beverage items printed on them. These became popular for use in drive-in restaurants and specialty houses. They are designed to speed service and reduce errors. These restaurants feature a specialty menu with the same food and beverage selection each day. Prices remain fixed for a fairly long period of time and

ZINO'S   7833

Server	Table No.	No. Guests	No. Checks

PIZZA	Large	Medium	Small	

D. D.	CLUB	STEAK	
SOLE	SUB.		
MTBL.	S.SUB		
TUNA	ZNVR.	F. F.	

G.P.	DU JOUR	ONION	
M. COAST	W. COAST		
TOSSED			

SP. SCE.	SP. MTBL.	
SP. MUSH.	LASAGNE	
V. PARM.		
	BROWN	
SEAFOOD	BEEF	
CHP. BF.	SPECIAL	

SOFT DRINK	LG	SM	
BEER	LG	SM	

	SUBTOTAL	
	TAX	
	TOTAL	

Fig. 14-3 Partly printed check.

BURGER CHEF

GO   STAY

NO. OF ITEMS	1	2	3	4	5	W	WO
Hamburgers	.70	1.40	2.10	2.80	3.50		
Funburgers	.80	1.60	2.40	3.20	4.00		
Cheeseburgers	.80	1.60	2.40	3.20	4.00		
Dbl. Cheeseburg.	1.40	2.80	4.20	5.60	7.00		
Skipper Trt.	1.30	2.60	3.90	5.20	6.50		
Big Chefs	1.50	3.00	4.50	6.00	7.50		
Funmeal	1.70	3.40	5.10	6.80	8.50		
Super Shefs	1.70	3.40	5.10	6.80	8.50		
Rancher							
Salad	.90	1.80	2.70	3.60	4.50		
Lg. Order Fries	.90	1.80	2.70	3.60	4.50		
French Fries	.60	1.20	1.80	2.40	3.00		
Turnovers	.50	1.00	1.50	2.00	2.50		
Cola - Lge	.60	1.20	1.80	2.40	3.00		
Cola - Reg.	.50	1.00	1.50	2.00	2.50		
Cola - Sm	.40	.80	1.20	1.60	2.00		
Orange - Lge	.60	1.20	1.80	2.40	3.00		
Orange - Reg.	.50	1.00	1.50	2.00	2.50		
Orange - Sm	.40	.80	1.20	1.60	2.00		
Root Beer - Lge	.60	1.20	1.80	2.40	3.00		
Root Beer - Reg.	.50	1.00	1.50	2.00	2.50		
Root Beer - Sm	.40	.80	1.20	1.60	2.00		
Lmnade - Lge	.60	1.20	1.80	2.40	3.00		
Lmnade - Reg.	.50	1.00	1.50	2.00	2.50		
Lmnade - Sm	.40	.80	1.20	1.60	2.00		
Van. Shake	.80	1.60	2.40	3.20	4.00		
Choc. Shake	.80	1.60	2.40	3.20	4.00		
Straw. Shake	.80	1.60	2.40	3.20	4.00		
Coffee	.40	.80	1.20	1.60	2.00		
Milk	.40	.80	1.20	1.60	2.00		
Hot Choc.	.40	.80	1.20	1.60	2.00		

N

SUBTOTAL _____

TAX _____

TOTAL _____

Fig. 14-4 Printed check.

usually are not changed until new checks are printed. The waiter, waitress, or carhop records any information requested at the top of the check, circles or underscores the selected items and their prices, extends prices if necessary, finds the subtotal, adds tax, and calculates the total. With the fluctuation of raw food prices in recent years, the printed check is declining in popularity.

SYSTEMS OF TAKING ORDERS

There are two systems used for taking the order and filling out the check.

1. The guest writes the order.

2. The waiter or waitress takes a verbal order from the guest.

Guest Writes Own Order

System No. 1 is usually used in private clubs, on trains, or in some in-plant food service operations. The guest is given a check and a pencil and instructed to write the desired order. It is not a very popular system. When this system is used, the following procedure is recommended.

1. Present the menu to the guest.

2. Make sure the information at the top of each check is completed.

3. Provide the guest with a check and a sharp pencil.

4. If the guest is hesitant to make out the check, offer assistance.

5. Your approach should be "Would you please write your own order? I will return for it shortly."

6. Collect the checks and go to the kitchen to place and assemble the orders.

7. When the food is served, enter the proper prices for each item listed. Extend the prices if there are two or more orders of the same item.

8. Total the check carefully to find the correct subtotal.

9. Enter the proper amount of tax. This is usually found by consulting a tax table.

10. Calculate the total and check all figures.

Waiter or Waitress Takes Verbal Order From Guest

In System No. 2 the waiter or waitress takes a verbal order from the guest, writes the order on a scratch pad, and later transfers the order to a sales check or writes the order directly on the sales check. Taking the order on a scratch pad first will usually produce a clearer, neater check. This system is by far the most popular because the guests feel they are getting more service. When this system is used the following procedure is recommended.

1. Present the menu to the guest.

2. Fill out the required information at the top of the check.

3. Enter on the check all items requested. Write neatly and clearly.

4. Take checks to the kitchen to place and assemble the order.

5. Enter the proper prices for each item listed. Extend prices correctly. Find the subtotal.

6. Enter the proper amount of tax. This is usually found by consulting a tax table. (Refer to figure 14-8, page 153.)

7. Calculate the total and check all figures.

Many food service establishments have different color checks. One is for breakfast, one for lunch, one for dinner, and still another color for room service or carryout orders. This is done to simplify record keeping and detect errors more rapidly.

A La Carte

Most better food service operations feature an *a la carte* menu (items priced separately) as well as a *table d'hote* menu (complete dinner). For a la carte orders, each item is priced separately. If the guest orders a shrimp cocktail, the price of the shrimp cocktail shown on the a la carte menu is recorded on the check. The guest may proceed to order a broiled lobster, julienne potatoes, salad, and coffee, and in each case a price for that particular item is recorded on the check. If a large amount of food is ordered, it can be quite expensive to order a la carte.

A la carte menus are preferred by certain guests—those who do not like the selections offered on the table d'hote menu, those who are on diets, or those who wish to order only one or two courses. Ordering from the table d'hote menu is the simplest and most economical way of dining out. The total price is shown opposite the entree (main course). This price usually includes appetizer, salad, potato, vegetable, entree, bread and butter, beverage, and in some cases, dessert. Most establishments charge extra for desserts and certain appetizers.

Stations

Each waiter or waitress is assigned a *station* (area of responsibility in the dining room, figure 14-5), consisting of a number of tables and chairs, booths, or section of a counter. Each station is numbered or sometimes each table is numbered.

Fig. 14-5 Waiter setting the tables on his station prior to opening for the luncheon business.

In either case, the waiter or waitress should be familiar with the area and numbers assigned. Placing these numbers on the check helps to eliminate confusion when being relieved or assisted by another member of the service crew. It also helps the waiter or waitress be more efficient and organized when serving.

Greeting Guests and Taking the Order

As the guests approach a service area, they should be greeted in an appropriate manner. Most establishments have their own standard procedure for greeting guests. Always smile and be cheerful. Assist the hostess or captain in seating the guests; never stand with arms folded. Check the table to be sure all necessary equipment is present and in its proper place. Fill the glasses with ice water while the hostess or captain presents the menu.

Standing to the left of the guest, with a book of checks and a sharp pencil, begin to fill in the required information at the top of the check. Then take the order according to the following steps.

1. Entree selection is given first. Ask how the item is to be cooked if this information applies to the selected item. Examples: roast beef, steaks, or chops; well done, medium, or rare. Notice whether the guests are ordering a la carte or table d'hote, since pricing will be different.

2. Appetizer selection. There may be an extra charge for this.

3. Salad selection and dressing desired. Name the dressings that are available.

4. Vegetable, if one is included, and potato selection. If a baked potato is selected, do they prefer butter or sour cream?

5. Beverage selection. If coffee is selected, would they like it served with the meal or with dessert?

6. Dessert selection. If this is not included on the table d'hote menu, an extra charge must be made.

Certain abbreviations are used by the waiter or waitress when taking an order to speed up the process. Some of the common abbreviations are given in figure 14-6.

Med.	Medium
R.	Rare
W.D.	Well Done
N.Y.	New York
T.I.	Thousand Island Dressing
Fr. D.	French Dressing
Sm.	Small
Lg.	Large
Din.	Dinner
Cof.	Coffee
Choc.	Chocolate
Fr. Fr.	French Fries
Van.	Vanilla
Tom.	Tomato
Ch.	Cheese
St. B.	String Beans

Fig. 14-6 Common abbreviations for taking orders.

Placing Order in the Kitchen

After the orders are taken, the waiter or waitress gathers the menus and goes to the preparation or kitchen area. The waiter or waitress is responsible for assembling juices, relishes, crackers, soups, bread, butter, condiments, and beverages. The entree (main course) is ordered from the cooking station responsible for that particular preparation. For example, broiled foods are ordered from the broiler cook; fried foods from the fry cook; stews, potted meats, and sautéed foods from the second cook; and so on. In smaller restaurants, all foods can be ordered and picked up in one area. When the foods are ready it is the responsibility of the waiter or waitress to pick them up and garnish them for service. Remember that food must first be ordered before it can be picked up. Often, a waiter or waitress will forget to place their orders and then arguments start and general confusion results.

After all foods are assembled and placed on a waiter's tray, the tray is taken to the expediter (food checker) who checks to see that only those items ordered by the guests are on the tray. This is done to control portions and food cost. The foods are then served to the guests in the proper order and fashion.

Totaling Check and Presenting to Guest

At the conclusion of the meal, start totaling the checks. Before presenting the checks to the guests, look for the following points:

1. Is all writing legible?

2. Are all items listed and priced properly?

3. Is the sales tax computed correctly?

4. Are subtotals and totals correct?

5. Is the information at the top of the check filled in correctly?

The check may be presented to the guest in two ways: (1) Under the edge of the guest's plate if single checks are given (if only one check for a group is given, place it under the host plate), or (2) Placed on a small tray, usually to the left of the guest. In both cases, the check is presented face down if payment is to be made to a cashier when the guest leaves, figure 14-7, or collected by the waiter or waitress at the end of the meal. If collection is to be made at the time the check is presented, it is then placed face up.

In many hotel or motel dining rooms, the guest can sign the check and have the amount added to the total bill. Payment is then made

Fig. 14-7 A guest presenting his check to the cashier.

when the guest checks out. In such cases, be sure the room number is marked on the check.

Minimum Charge

Establishments featuring live music, a floor show, or some special type of entertainment usually add a cover charge to the check. The cover charge is a form of admission fee charged to each person to help pay for the cost of the entertainment. Another method of collecting for entertainment or service is by having a *minimum charge*. This means that a guest is required to spend a certain amount of money, once seated, even if the total check amounts to less. For example, if the minimum charge is $5.00, but the check amounts to only $4.25, the guest is still required to pay the $5.00 minimum charge.

Sales Tax

Sales tax, which is discussed in another part of the book, is important in making out the sales check correctly. Most states have a sales tax; the rate varies depending on the state. Tax tables are made available so the amount of tax can be calculated quickly and correctly. An example of a sales tax collection table is shown in figure 14-8. Use it in completing the exercise problems.

SALES 15¢ AND UNDER - NO TAX

Transaction ... Tax	Transaction ... Tax	Transaction ... Tax
.16 to .22 - .01	16.67 to 16.88 - .76	33.34 to 33.55 - 1.51
.23 to .44 - .02	16.89 to 17.11 - .77	33.56 to 33.77 - 1.52
.45 to .66 - .03	17.12 to 17.33 - .78	33.78 to 34.00 - 1.53
.67 to .88 - .04	17.34 to 17.55 - .79	34.01 to 34.22 - 1.54
.89 to 1.11 - .05	17.56 to 17.77 - .80	34.23 to 34.44 - 1.55
1.12 to 1.33 - .06	17.78 to 18.00 - .81	34.45 to 34.66 - 1.56
1.34 to 1.55 - .07	18.01 to 18.22 - .82	34.67 to 34.88 - 1.57
1.56 to 1.77 - .08	18.23 to 18.44 - .83	34.89 to 35.11 - 1.58
1.78 to 2.00 - .09	18.45 to 18.66 - .84	35.12 to 35.33 - 1.59
2.01 to 2.22 - .10	18.67 to 18.88 - .85	35.34 to 35.55 - 1.60
2.23 to 2.44 - .11	18.89 to 19.11 - .86	35.56 to 35.77 - 1.61
2.45 to 2.66 - .12	19.12 to 19.33 - .87	35.78 to 36.00 - 1.62
2.67 to 2.88 - .13	19.34 to 19.55 - .88	36.01 to 36.22 - 1.63
2.89 to 3.11 - .14	19.56 to 19.77 - .89	36.23 to 36.44 - 1.64
3.12 to 3.33 - .15	19.78 to 20.00 - .90	36.45 to 36.66 - 1.65
3.34 to 3.55 - .16	20.01 to 20.22 - .91	36.67 to 36.88 - 1.66
3.56 to 3.77 - .17	20.23 to 20.44 - .92	36.89 to 37.11 - 1.67
3.78 to 4.00 - .18	20.45 to 20.66 - .93	37.12 to 37.33 - 1.68
4.01 to 4.22 - .19	20.67 to 20.88 - .94	37.34 to 37.55 - 1.69
4.23 to 4.44 - .20	20.89 to 21.11 - .95	37.56 to 37.77 - 1.70
4.45 to 4.66 - .21	21.12 to 21.33 - .96	37.78 to 38.00 - 1.71
4.67 to 4.88 - .22	21.34 to 21.55 - .97	38.01 to 38.22 - 1.72
4.89 to 5.11 - .23	21.56 to 21.77 - .98	38.23 to 38.44 - 1.73
5.12 to 5.33 - .24	21.78 to 22.00 - .99	38.45 to 38.66 - 1.74
5.34 to 5.55 - .25	22.01 to 22.22 - 1.00	38.67 to 38.88 - 1.75
5.56 to 5.77 - .26	22.23 to 22.44 - 1.01	38.89 to 39.11 - 1.76
5.78 to 6.00 - .27	22.45 to 22.66 - 1.02	39.12 to 39.33 - 1.77
6.01 to 6.22 - .28	22.67 to 22.88 - 1.03	39.34 to 39.55 - 1.78
6.23 to 6.44 - .29	22.89 to 23.11 - 1.04	39.56 to 39.77 - 1.79
6.45 to 6.66 - .30	23.12 to 23.33 - 1.05	39.78 to 40.00 - 1.80
6.67 to 6.88 - .31	23.34 to 23.55 - 1.06	40.01 to 40.22 - 1.81
6.89 to 7.11 - .32	23.56 to 23.77 - 1.07	40.23 to 40.44 - 1.82
7.12 to 7.33 - .33	23.78 to 24.00 - 1.08	40.45 to 40.66 - 1.83
7.34 to 7.55 - .34	24.01 to 24.22 - 1.09	40.67 to 40.88 - 1.84
7.56 to 7.77 - .35	24.23 to 24.44 - 1.10	40.89 to 41.11 - 1.85
7.78 to 8.00 - .36	24.45 to 24.66 - 1.11	41.42 to 41.33 - 1.86
8.01 to 8.22 - .37	24.67 to 24.88 - 1.12	41.34 to 41.55 - 1.87
8.23 to 8.44 - .38	24.89 to 25.11 - 1.13	41.56 to 41.77 - 1.88
8.45 to 8.66 - .39	25.12 to 25.33 - 1.14	41.78 to 42.00 - 1.89
8.67 to 8.88 - .40	25.34 to 25.55 - 1.15	42.01 to 42.22 - 1.90
8.89 to 9.11 - .41	25.56 to 25.77 - 1.16	42.23 to 42.44 - 1.91
9.12 to 9.33 - .42	25.78 to 26.00 - 1.17	42.45 to 42.66 - 1.92
9.34 to 9.55 - .43	26.01 to 26.22 - 1.18	42.67 to 42.88 - 1.93
9.56 to 9.77 - .44	26.23 to 26.44 - 1.19	42.89 to 43.11 - 1.94
9.78 to 10.00 - .45	26.45 to 26.66 - 1.20	43.12 to 43.33 - 1.95
10.01 to 10.22 - .46	26.67 to 26.88 - 1.21	43.34 to 43.55 - 1.96
10.23 to 10.44 - .47	26.89 to 27.11 - 1.22	43.56 to 43.77 - 1.97
10.45 to 10.66 - .48	27.12 to 27.33 - 1.23	43.78 to 44.00 - 1.98
10.67 to 10.88 - .49	27.34 to 27.55 - 1.24	44.01 to 44.22 - 1.99
10.89 to 11.11 - .50	27.56 to 27.77 - 1.25	44.23 to 44.44 - 2.00
11.12 to 11.33 - .51	27.78 to 28.00 - 1.26	44.45 to 44.66 - 2.01
11.34 to 11.55 - .52	28.01 to 28.22 - 1.27	44.67 to 44.88 - 2.02
11.56 to 11.77 - .53	28.23 to 28.44 - 1.28	44.89 to 45.11 - 2.03
11.78 to 12.00 - .54	28.45 to 28.66 - 1.29	45.12 to 45.33 - 2.04
12.01 to 12.22 - .55	28.67 to 28.88 - 1.30	45.34 to 45.55 - 2.05
12.23 to 12.44 - .56	28.89 to 29.11 - 1.31	45.56 to 45.77 - 2.06
12.45 to 12.66 - .57	29.12 to 29.33 - 1.32	45.78 to 46.00 - 2.07
12.67 to 12.88 - .58	29.34 to 29.55 - 1.33	46.01 to 46.22 - 2.08
12.89 to 13.11 - .59	29.56 to 29.77 - 1.34	46.23 to 46.44 - 2.09
13.12 to 13.33 - .60	29.78 to 30.00 - 1.35	46.45 to 46.66 - 2.10
13.34 to 13.55 - .61	30.01 to 30.22 - 1.36	46.67 to 46.88 - 2.11
13.56 to 13.77 - .62	30.23 to 30.44 - 1.37	46.89 to 47.11 - 2.12
13.78 to 14.00 - .63	30.45 to 30.66 - 1.38	47.12 to 47.33 - 2.13
14.01 to 14.22 - .64	30.67 to 30.88 - 1.39	47.34 to 47.55 - 2.14
14.23 to 14.44 - .65	30.89 to 31.11 - 1.40	47.56 to 47.77 - 2.15
14.45 to 14.66 - .66	31.12 to 31.33 - 1.41	47.78 to 48.00 - 2.16
14.67 to 14.88 - .67	31.34 to 31.55 - 1.42	48.01 to 48.22 - 2.17
14.89 to 15.11 - .68	31.56 to 31.77 - 1.43	48.23 to 48.44 - 2.18
15.12 to 15.33 - .69	31.78 to 32.00 - 1.44	48.45 to 48.66 - 2.19
15.34 to 15.55 - .70	32.01 to 32.22 - 1.45	48.67 to 48.88 - 2.20
15.56 to 15.77 - .71	32.23 to 32.44 - 1.46	48.89 to 49.11 - 2.21
15.78 to 16.00 - .72	32.45 to 32.66 - 1.47	49.12 to 49.33 - 2.22
16.01 to 16.22 - .73	32.67 to 32.88 - 1.48	49.34 to 49.55 - 2.23
16.23 to 16.44 - .74	32.89 to 33.11 - 1.49	49.56 to 49.77 - 2.24
16.45 to 16.66 - .75	33.12 to 33.33 - 1.50	49.78 to 50.00 - 2.25

Fig. 14-8 Sales tax table

—ACHIEVEMENT REVIEW—
SALES TAX AND TOTALING CHECKS

Complete exercises 1 to 10 by filling in the subtotal, tax, and total. The tax can be found by referring to the sales tax table in figure 14-8.

1.

ZINO'S   7833

Server	Table No.	No. Guests	No. Checks
BH	7	5	1

PIZZA	Large	Medium	Small	
	1			6.00
anchovy				

D. D.	CLUB	STEAK	
SOLE	SUB.		
MTBL.	S.SUB 1		2.60
TUNA 1	ZNVR.	F. F.	2.30

G.P.	DU JOUR	ONION	
M. COAST	W. COAST		
TOSSED			

SP. SCE.		SP. MTBL.	
SP. MUSH. 1		LASAGNE	2.50
V. PARM.			
		BROWN	
SEAFOOD		BEEF	
CHP. BF.		SPECIAL	

SOFT DRINK	3	LG	SM	1.80
BEER	2	LG	SM	1.40

	SUBTOTAL	
	TAX	
	TOTAL	

2.

ZINO'S   7833

Server	Table No.	No. Guests	No. Checks
BH	8	6	1

PIZZA	Large	Medium	Small	
	2			10.50
Pepperoni				

D. D.	CLUB	STEAK 1	5.50
SOLE	SUB.		
MTBL.	S.SUB		
TUNA	ZNVR.	F. F.	

G.P.	DU JOUR	ONION	
M. COAST	W. COAST		
TOSSED			

SP. SCE.		SP. MTBL. 1	2.40
SP. MUSH.		LASAGNE 2	3.20
V. PARM.			
		BROWN	
SEAFOOD		BEEF	
CHP. BF.		SPECIAL	

SOFT DRINK	2	LG	SM	1.20
BEER	4	LG	SM	2.80

	SUBTOTAL	
	TAX	
	TOTAL	

3.

ZINO'S

7833

Server	Table No.	No. Guests	No. Checks
BH	6	4	1

PIZZA	Large	Medium	Small	
1				5.50
mush.				

D.D.	CLUB 1	STEAK	3.00
SOLE	SUB.		5.20
MTBL.	S.SUB		
TUNA	ZNVR.	F.F.	1.40

G.P.	DU JOUR	ONION	1.60
M. COAST	W. COAST		
TOSSED			1.80

SP. SCE.	SP. MTBL.	
SP. MUSH.	LASAGNE	
V. PARM.		
	BROWN	
SEAFOOD	BEEF	
CHP. BF.	SPECIAL	

SOFT DRINK	2	LG	SM	1.20
BEER	2	LG	SM	1.40

	SUBTOTAL	
	TAX	
	TOTAL	

4.

LaRosa's

Date 7/18	Server CW	Table No. 8	No. Persons 4	No. 487176

1	Sirloin Steak m/w	
2	B. Pot. Gr. B.	12.00
3	Sl. Tom. Fr. O	
4		
5	Broiled Steak Trout	
6	Hash Br. Peas	10.50
7	Toss Salad T.I.	
8		
9	Sauteed Calf's Liver	
10	Hash Br. Corn	9.00
11	Toss Salad	
12		
13	Fr. Chick	
14	M. Pot. Peas	7.50
15	Toss Salad T.I.	
16		
17	2 Cof w cr.	
18	2 Ice Tea	
19	4 apple pie	3.20
20		
21		
22	SUBTOTAL	
23	TAX	
24	TOTAL	
—		No. 487176

Date	Guest Receipt	Persons	Amount of Check	

LaRosa's

5.

LaRosa's

Date 7/18	Server CW	Table No. 7	No. Persons 3	NO. 487176
1				
2	Shrimp Ct.			3.00
3	T. Bone med.			12.50
4	Fr. Fr. Pot. Lima B			
5	Cole Slaw			
6				
7	Seafood Ct.			2.50
8	Club Steak med			10.50
9	Hash Br St B			
10	Sl Tom Sal Fr O			
11				
12	Crabmeat Ct			3.50
13	Filet Mignon R			14.00
14				
15	1 Hot Tea			
16	2 Cof w Cr			
17				
18	3 Straw Short Cake			4.50
19				
20				
21				
22		SUBTOTAL		
23		TAX		
24		TOTAL		
—				NO. 487176

Date	Guest Receipt	Persons	Amount of Check	

 LaRosa's

6.

LaRosa's

Date 7/18	Server CW	Table No. 9	No. Persons 2	NO. 487176
1	a la carte			
2	Shrimp Ct.			3.00
3	Broiled Calf's Liver med			6.50
4	Cottage Fr. Pot.			1.30
5	Stewed Tom.			.90
6	Coffee w cr.			.60
7				
8				
9	Broiled Grapefruit			1.50
10	Broiled Lobster			13.00
11	Julienne Pot.			1.30
12	Toss Salad T.I.			1.50
13	Ice Tea			.60
14	apple Pie			.90
15				
16				
17				
18				
19				
20				
21				
22		SUBTOTAL		
23		TAX		
24		TOTAL		
—				NO. 487176

Date	Guest Receipt	Persons	Amount of Check	

LaRosa's

7.

LaRosa's

Date 7/18	Server CW	Table No. 10	No. Persons 3	No. 487176

1		*a la carte*	
2	Sautéed Trout	7.00
3	Cottage Fr. Pot.	1.30
4	Chef Salad T.I.	1.50
5	Coffee w Cr.	.60
6		
7	Fr. Froglegs	6.50
8	Hash Br. Pot.	1.30
9	Chef Salad T.I.	1.50
10	Ice Tea	.60
11		
12	Shrimp Ct.	3.00
13	Fr. Chick.	5.50
14	Fr. Fr. Pot.	1.50
15	Toss Sal. Fr. D.	1.30
16	Milk	.60
17		
18	3 cherry pie	2.70
19		
20		
21		
22	SUBTOTAL	
23	TAX	
24	TOTAL	
—		No. 487176

Date	Guest Receipt	Persons	Amount of Check	

LaRosa's

8.

		BURGER CHEF					
GO ✓ STAY							
NO. OF ITEMS	**1**	**2**	**3**	**4**	**5**	**W**	**WO**
Hamburgers	.70	1.40	2.10	2.80	3.50		
Funburgers	.80	1.60	2.40	3.20	4.00		
Cheeseburgers	.80	1.60	2.40	3.20	(4.00)		
Dbl. Cheeseburg.	1.40	2.80	4.20	5.60	7.00		
Skipper Trt.	1.30	2.60	3.90	5.20	6.50		
Big Chefs	1.50	3.00	4.50	6.00	7.50		
Funmeal	1.70	3.40	5.10	6.80	8.50		
Super Shefs	1.70	3.40	5.10	6.80	8.50		
Rancher							
Salad	.90	1.80	2.70	3.60	4.50		
Lg. Order Fries	.90	1.80	2.70	3.60	4.50		
French Fries	.60	1.20	1.80	2.40	(3.00)		
Turnovers	.50	1.00	1.50	2.00	(2.50)		
Cola - Lge	.60	1.20	1.80	2.40	3.00		
Cola - Reg.	.50	1.00	1.50	2.00	2.50		
Cola - Sm	.40	.80	1.20	1.60	2.00		
Orange - Lge	.60	1.20	1.80	2.40	3.00		
Orange - Reg.	.50	1.00	1.50	2.00	2.50		
Orange - Sm	.40	.80	1.20	1.60	2.00		
Root Beer - Lge	.60	1.20	1.80	2.40	3.00		
Root Beer - Reg.	.50	1.00	1.50	2.00	2.50		
Root Beer - Sm	.40	.80	1.20	1.60	2.00		
Lmnade - Lge	.60	1.20	1.80	2.40	3.00		
Lmnade - Reg.	.50	1.00	1.50	2.00	2.50		
Lmnade - Sm	.40	.80	1.20	1.60	2.00		
Van. Shake	.80	1.60	2.40	3.20	(4.00)		
Choc. Shake	.80	1.60	2.40	3.20	4.00		
Straw. Shake	.80	1.60	2.40	3.20	4.00		
Coffee	.40	.80	1.20	1.60	2.00		
Milk	.40	.80	1.20	1.60	2.00		
Hot Choc.	.40	.80	1.20	1.60	2.00		

N

SUBTOTAL _____
TAX _____
TOTAL _____

9.

GO ✓ STAY — BURGER CHEF							
NO. OF ITEMS	1	2	3	4	5	W	WO
Hamburgers	.70	1.40	(2.10)	2.80	3.50		
Funburgers	.80	1.60	2.40	3.20	4.00		
Cheeseburgers	.80	1.60	2.40	3.20	4.00		
Dbl. Cheeseburg.	1.40	2.80	4.20	5.60	7.00		
Skipper Trt.	1.30	2.60	3.90	5.20	6.50		
Big Chefs	1.50	(3.00)	4.50	6.00	7.50		
Funmeal	1.70	3.40	5.10	6.80	8.50		
Super Shefs	1.70	3.40	5.10	6.80	8.50		
Rancher							
Salad	.90	1.80	2.70	3.60	4.50		
Lg. Order Fries	.90	1.80	2.70	3.60	4.50		
French Fries	.60	1.20	1.80	2.40	(3.00)		
Turnovers	.50	1.00	1.50	2.00	2.50		
Cola - Lge	.60	1.20	1.80	2.40	3.00		
Cola - Reg.	.50	(1.00)	1.50	2.00	2.50		
Cola - Sm	.40	.80	1.20	1.60	2.00		
Orange - Lge	.60	1.20	1.80	2.40	3.00		
Orange - Reg.	.50	1.00	1.50	2.00	2.50		
Orange - Sm	.40	.80	1.20	1.60	2.00		
Root Beer - Lge	(.60)	1.20	1.80	2.40	3.00		
Root Beer - Reg.	.50	1.00	1.50	2.00	2.50		
Root Beer - Sm	.40	.80	1.20	1.60	2.00		
Lmnade - Lge	.60	(1.20)	1.80	2.40	3.00		
Lmnade - Reg.	.50	1.00	1.50	2.00	2.50		
Lmnade - Sm	.40	.80	1.20	1.60	2.00		
Van. Shake	.80	1.60	2.40	3.20	4.00		
Choc. Shake	.80	1.60	2.40	3.20	4.00		
Straw. Shake	.80	1.60	2.40	3.20	4.00		
Coffee	.40	.80	1.20	1.60	2.00		
Milk	.40	.80	1.20	1.60	2.00		
Hot Choc.	.40	.80	1.20	1.60	2.00		

N

SUBTOTAL _____
TAX _____
TOTAL _____

10.

GO STAY ✓ — BURGER CHEF							
NO. OF ITEMS	1	2	3	4	5	W	WO
Hamburgers	.70	1.40	2.10	2.80	3.50		
Funburgers	.80	1.60	2.40	3.20	4.00		
Cheeseburgers	.80	1.60	2.40	3.20	4.00		
Dbl. Cheeseburg.	1.40	(2.80)	4.20	5.60	7.00		
Skipper Trt.	1.30	2.60	3.90	5.20	6.50		
Big Chefs	1.50	3.00	4.50	6.00	7.50		
Funmeal	1.70	3.40	5.10	6.80	8.50		
Super Shefs	1.70	(3.40)	5.10	6.80	8.50		
Rancher							
Salad	.90	1.80	(2.70)	3.60	4.50		
Lg. Order Fries	.90	1.80	2.70	3.60	4.50		
French Fries	.60	1.20	1.80	(2.40)	3.00		
Turnovers	.50	1.00	1.50	2.00	2.50		
Cola - Lge	(.60)	1.20	1.80	2.40	3.00		
Cola - Reg.	.50	1.00	1.50	2.00	2.50		
Cola - Sm	.40	.80	1.20	1.60	2.00		
Orange - Lge	.60	1.20	(1.80)	2.40	3.00		
Orange - Reg.	.50	1.00	1.50	2.00	2.50		
Orange - Sm	.40	.80	1.20	1.60	2.00		
Root Beer - Lge	.60	1.20	1.80	2.40	3.00		
Root Beer - Reg.	(.50)	1.00	1.50	2.00	2.50		
Root Beer - Sm	.40	.80	1.20	1.60	2.00		
Lmnade - Lge	.60	1.20	1.80	2.40	3.00		
Lmnade - Reg.	.50	1.00	1.50	2.00	2.50		
Lmnade - Sm	.40	.80	1.20	1.60	2.00		
Van. Shake	.80	1.60	2.40	3.20	4.00		
Choc. Shake	.80	1.60	2.40	3.20	4.00		
Straw. Shake	.80	1.60	2.40	3.20	4.00		
Coffee	.40	.80	1.20	1.60	2.00		
Milk	.40	.80	1.20	1.60	2.00		
Hot Choc.	.40	.80	1.20	1.60	2.00		

N

SUBTOTAL _____
TAX _____
TOTAL _____

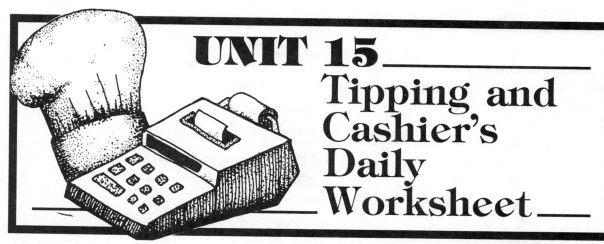

UNIT 15
Tipping and Cashier's Daily Worksheet

TIPPING

Tipping, also referred to as a gratuity, is the giving of a small fee for a service rendered. It is used in the food service industry as a reward for courteous, prompt service.

Some people disagree with this procedure and regardless of the quality of the service, refuse to tip. They assume that the waiter or waitress is paid an adequate salary by the employer to take the order and serve the food. Other people do not feel obligated to tip, but will leave a tip if courteous, prompt service is received. Therefore, the amount of tip received by a waiter or waitress usually depends on the attitude of the guest towards tipping, also whether the meal was satisfying, and certainly the quality of service received, figure 15-1.

Most people tip at a percentage of the check. The accepted practice is to tip 15 percent of the amount of the check. Some establishments automatically add 15 percent of the check amount to the bill. If this is done, the guests should be made aware of this policy before they are served. Often, when the 15 percent is added, the guest is not aware of this policy, and, therefore, still leaves a tip at the table.

For the guest who wishes to tip 15 percent of the amount of the check, or for the waiter or waitress who is asked to figure what 15 percent of the check is, there is an easy way to calculate the amount. First, find 10 percent of the bill by

Fig. 15-1 The tip is the guest's way of showing his appreciation for excellent service.

moving the decimal point in the total one place to the left. Next, take half of the figure just found. Then add the two figures together for the tip. Example: the bill is $18.00. 10 percent of $18.00 is $1.80. Half of $1.80 is $.90. Add the two together: $1.80 + $.90 = $2.70, the amount of the tip.

ACHIEVEMENT REVIEW
TIPPING

Find the amount of tip for each of the following bills if the tip equals 15 percent of the bill. Do not round off the amount.

1.	$12.00	11.	$50.95
2.	$15.00	12.	$52.85
3.	$20.00	13.	$60.95
4.	$18.80	14.	$70.65
5.	$24.25	15.	$80.45
6.	$25.40	16.	$82.60
7.	$30.25	17.	$95.85
8.	$35.60	18.	$105.40
9.	$32.50	19.	$110.80
10.	$40.45	20.	$125.50

CASHIER'S DAILY WORKSHEET

The cashier's daily worksheet is another control method used by management to keep track of cash sales. Its purpose is to determine whether the actual amount of cash in the register drawer equals the total amount of cash sales made during a specific time period. The report may show a very small amount over or under; with the constant exchange of cash between the guests and the cashier, small mistakes may occur. Usually, management only becomes concerned when these amounts exceed a dollar. The worksheet is designed not only to protect the business operator from theft, but also to protect the cashier. If the cashier makes a costly mistake, the error can usually be found by checking the daily worksheet.

There are many different types of cashier's daily worksheets in use because most establishments create their own form that is best suited for their particular operation. All forms contain the same general information. An example of a typical worksheet is given in figure 15-2, followed by a step-by-step explanation of how it is completed.

Explanation of Items on Worksheet

Receipts (Register Readings). These figures are taken from the cash register. With the versatile

```
                    Cashier's Daily Worksheet
    Date _____
    Receipts (Register Readings)
          Food  $450.00
          Liquor $290.00
          Misc. Items $120.00
          Sales Tax  $38.70
    Gross Receipts                              $898.70
    Add—Change                                  $ 25.00
    Total Cash                                  $923.70
    Less—Cash Paid Outs                         $ 11.65
    Total Cash in Drawer                        $912.05
    Cash—Actual                                 $911.65
    Over—Short                                  $   .40
                    Record of Cash Paid Outs
    To Whom Paid                                Amount
    City Ice Company                            $  2.90
    Johnson Sharpening Service                  $  3.50
    Watson Florist                              $  5.25

                          Total Cash Paid Out  $ 11.65
    Weather Cloudy and Cool
    Customer Count 210
                                   Signed Jane Carson
```

Fig. 15-2 Cashier's daily worksheet

cash registers in use today, figure 15-3, items can be categorized, rung up, and totaled separately or totaled together.

Gross Receipts. A total of all the separate register readings. The gross is the total before any deductions are made.

Add—Change. Starting change is added to the gross receipts. This money is placed in the register before any sales are made.

Total Cash. This figure represents the amount of cash that should be in the cash drawer before any paid outs are made.

Less—Cash Paid Outs. This figure is acquired by totaling the amounts of money paid out of the cash register during the day. When a paid out occurs, a record must be made of the transaction by recording it on the worksheet in the section headed Record of Cash Paid Outs. For each cash paid out, the cashier should have a receipted bill, invoice, or cash payment voucher. Most registers have a key for recording paid outs. The total amount of paid outs on the worksheet should equal the total amount of paid outs recorded by the register. Paid outs are subtracted from the total cash because this money was taken out of the register drawer.

Total Cash in Drawer. Total cash, less paid

Fig. 15-3 Cashier in a fast food service operation ringing up a sale.

outs, gives the amount of cash that should be in the cash register drawer at the end of the day.

Cash—Actual. The amount of cash that is actually in the cash register drawer after all coins, currency, and checks are totaled.

Over—Short. If the amounts shown in Total Cash in Drawer and Cash-Actual are not equal, the difference is recorded as cash over or short. If the amount shown in Total Cash in Drawer is more than cash-actual, there is a cash shortage. If the Cash-Actual is more than the Total Cash in Drawer, there is a cash surplus.

Record of Cash Paid Outs. All money paid out of the register drawer is recorded here with the name of the person or company to whom it was paid.

Weather. A record is kept of the weather for each day because weather may influence the amount of business done.

Customer Count. This figure is produced by the cash register. The cash register records the number of customers served as each sale is rung up on the register.

Signed. The cashier checks all figures and is then required to sign the worksheet.

ACHIEVEMENT REVIEW
CASHIER'S DAILY WORKSHEET

Complete the following cashier's daily worksheets, as shown in the example in figure 15-2.

1.

Date _____

Receipts (Register Readings)

      Food <u>$585.54</u>

      Liquor <u>$392.25</u>

      Misc. Items <u>$94.74</u>

      Sales Tax <u>$48.26</u>

Gross Receipts

Add—Change $ 50.00

Total Cash

Less—Cash Paid Outs

Total Cash in Drawer

Cash—Actual $1131.23

Over—Short

Record of Cash Paid Outs

To *Whom* Paid   Amount

Smith Candy Co. $ 12.93

Watson Florist $ 15.24

City Ice Co. $ 8.79

Weather *Cloudy and Warm*   Total Cash Paid Out _____

Customer Count *312*

                     Signed _____

2.

```
Date _____
Receipts (Register Readings)
     Food $785.92
     Liquor $496.83
     Misc. Items $83.26
     Sales Tax $61.47
Gross Receipts
Add—Change                              $   45.00
Total Cash
Less—Cash Paid Outs
Total Cash in Drawer
Cash—Actual                             $1429.28
Over—Short

              Record of Cash Paid Outs
To Whom Paid                            Amount
Johnson Sharpening Service              $   20.15
City Ice Co.                            $   15.82


                         Total Cash Paid Out   _____
Weather Clear and Sunny
Customer Count 250
                         Signed _____
```

3.

```
Date _____
Receipts (Register Readings)
     Food $597.20
     Liquor $695.80
     Misc. Items $76.43
     Sales Tax $61.62
Gross Receipts
Add—Change                              $   35.00
Total Cash
Less—Cash Paid Outs
Total Cash in Drawer
Cash—Actual                             $1429.43
Over—Short

              Record of Cash Paid Outs
To Whom Paid                            Amount
City Ice Co.                            $   10.85
Life Uniform Co.                        $   15.93
Watson Florist                          $   12.36


                         Total Cash Paid Out   _____
Weather Cloudy and Cool
Customer Count 290
                         Signed _____
```

4.

Date _____

Receipts (Register Readings)

    Food <u>$1275.65</u>

    Liquor <u>$824.30</u>

    Misc. Items <u>$116.45</u>

    Sales Tax <u>$99.74</u>

Gross Receipts

Add—Change   <u>$ 75.00</u>

Total Cash

Less—Cash Paid Outs   <u>   </u>

Total Cash in Drawer   <u>   </u>

Cash—Actual   <u>$2351.54</u>

Over—Short   <u>   </u>

Record of Cash Paid Outs

To *Whom* Paid   <u>Amount</u>

Jones Candle Co.   <u>$ 13.92</u>

City Ice Co.   <u>$ 15.73</u>

Watson Florist   <u>$ 8.76</u>

                                <u>   </u>

Weather *Clear and Cool*   Total Cash Paid Out <u>   </u>

Customer Count *420*

                   Signed _____

5.

Date _____

Receipts (Register Readings)

    Food <u>$856.63</u>

    Liquor <u>$474.68</u>

    Misc. Items <u>$91.21</u>

    Sales Tax <u>$64.01</u>

Gross Receipts

Add—Change   <u>$ 25.00</u>

Total Cash

Less—Cash Paid Outs   <u>   </u>

Total Cash in Drawer   <u>   </u>

Cash—Actual   <u>$1454.43</u>

Over—Short   <u>   </u>

Record of Cash Paid Outs

To *Whom* Paid   <u>Amount</u>

City Ice Co.   <u>$ 21.50</u>

Smith Candy Co.   <u>$ 16.85</u>

Watson Florist   <u>$ 18.75</u>

                                <u>   </u>

Weather *Rain*   Total Cash Paid Out <u>   </u>

Customer Count *275*

                   Signed _____

6.

Date _____	
Receipts (Register Readings)	
Food $1286.64	
Liquor $728.58	
Misc. Items $118.92	
Sales Tax $96.04	
Gross Receipts	_____
Add—Change	$ 65.00
Total Cash	_____
Less—Cash Paid Outs	_____
Total Cash in Drawer	_____
Cash—Actual	$2243.67
Over—Short	_____

Record of Cash Paid Outs

To *Whom* Paid	Amount
Life Uniform Co.	$ 23.45
City Ice Co.	$ 15.50
Watson Florist	$ 8.65
Smith Candy Co.	$ 7.60
Total Cash Paid Out	_____

Weather *Clear and Cold*

Customer Count *350*

Signed _____

7.

Date _____	
Receipts (Register Readings)	
Food $926.45	
Liquor $438.35	
Misc. Items $67.25	
Sales Tax $64.44	
Gross Receipts	_____
Add—Change	$ 55.00
Total Cash	_____
Less—Cash Paid Outs	_____
Total Cash in Drawer	_____
Cash—Actual	$1526.68
Over—Short	_____

Record of Cash Paid Outs

To *Whom* Paid	Amount
City Ice Co.	$ 13.35
Smith Candy Co.	$ 11.46
Total Cash Paid Out	_____

Weather *Clear and Cool*

Customer Count *405*

Signed _____

8.

Date _____

Receipts (Register Readings)

Food $1565.00

Liquor $875.93

Misc. Items $78.68

Sales Tax $113.38

Gross Receipts

Add—Change ... $ 60.00

Total Cash

Less—Cash Paid Outs

Total Cash in Drawer

Cash—Actual ... $2627.31

Over—Short

Record of Cash Paid Outs

To *Whom* Paid ... Amount

Smith Candy Co. ... $ 13.63

Life Uniform Co. ... $ 17.67

National Linen Co. ... $ 26.43

Total Cash Paid Out _____

Weather *Snow and Cold*

Customer Count *520*

Signed _____

9.

Date _____

Receipts (Register Readings)

Food $758.75

Liquor $396.85

Misc. Items $58.25

Sales Tax $54.62

Gross Receipts

Add—Change ... $ 45.00

Total Cash

Less—Cash Paid Outs

Total Cash in Drawer

Cash—Actual ... $1272.83

Over—Short

Record of Cash Paid Outs

To *Whom* Paid ... Amount

National Linen Co. ... $ 28.38

Watson Florist ... $ 15.26

Total Cash Paid Out _____

Weather *Cloudy and Cold*

Customer Count *325*

Signed _____

10.

Date _____	
Receipts (Register Readings)	
Food <u>$2589.68</u>	
Liquor <u>$1648.23</u>	
Misc. Items <u>$362.60</u>	
Sales Tax <u>$207.02</u>	
Gross Receipts	
Add—Change	$ 75.00
Total Cash	
Less—Cash Paid Outs	
Total Cash in Drawer	
Cash—Actual	<u>$4616.89</u>
Over—Short	

Record of Cash Paid Outs

To *Whom* Paid	Amount
Watson Florist	$ 18.85
City Ice Co.	$ 16.25
National Linen Co.	$ 25.40
Total Cash Paid Out	_____

Weather *Fair and Hot*
Customer Count *725*

Signed _____

PART FOUR

Management— Operational Procedures

The function of management in a food service operation is to direct people, provide efficient service, and control both money and materials so a profit can be made. In this section, the emphasis is on the operational procedures that help management control money and materials, and at the same time provide the records necessary for a good accounting system.

Not all food service students have the desire or ability to manage a food service establishment. However, it is helpful to learn management procedures to better understand the functions of management and to know what makes a successful operation. With this knowledge, the student can become a better food service employee, which may lead to a more responsible position.

The operational procedures that are important to management in controlling money and materials are as follows:

- Figuring standard recipe cost
- Pricing the menu
- Daily food cost report
- Perpetual and physical inventory
- Financial statements
- Break-even analysis
- Budgeting

UNIT 16
Figuring Standard Recipe Cost

In any business venture, it is necessary to know the cost of the items to be sold before establishing a selling price. This is why it is very important to standardize and cost all recipes used in a food service establishment.

When figuring the cost of a standard recipe, the cost of each ingredient that goes into the preparation is totaled and a per-unit cost is calculated. For example, when preparing a certain recipe that yields 75 servings, the costs of all the ingredients used in making these 75 servings are added up to obtain the complete cost of the preparation. The total cost of the preparation is then divided by 75 (the yield) to find the unit cost. The unit cost represents what one serving of this particular item costs to prepare. With this knowledge, the manager or food and beverage controller can establish a menu price.

Prices of most ingredients fluctuate (go up or down) from time to time. Therefore, the manager or food and beverage controller must be alert to any price changes and make the necessary adjustments on the unit cost of the recipes as these situations occur.

The forms used to record the cost of a certain recipe may differ depending on the food service establishment, but they all usually supply the same information—the exact cost to produce one serving. Of course, in order to obtain the proper yield and determine unit cost, all servings must be uniform in size, figure 16-1, page 172. A typical standardized recipe showing the market price of each ingredient, the extension cost, total cost, and unit cost per serving is given in figure 16-2.

- The first column lists all the ingredients used in the preparation, in this case, Hungarian Beef Goulash.
- The second column lists the amount of each ingredient needed to prepare 50—6-ounce servings.
- The third column lists the current market price of each ingredient. The price is listed in quantities that are usually quoted by the purveyor. For example, meat is always purchased by the pound, eggs by the dozen, milk by the gallon, and so forth.
- The fourth column is the extension cost of the quantity listed. This figure is found by multiplying the amount by the price. Remember when multiplying, both figures

Fig. 16-1 All rolls must be uniform in size in order to obtain the proper yield and determine unit cost.

should represent the same quantity. That is, pounds multiplied by price per pound, ounces multiplied by price per ounce, and so on. Total the extension column to find the total cost of the preparation.

• To find the cost per serving, divide the total cost by the number of servings the recipe will yield. In this case, the yield is 50, so 50 is divided into $26.01 to get $.52, which is the cost of each serving. Note: Prices have been carried to the tenth of a cent (mill), three places to the right of the decimal.

Hungarian Beef Goulash		Approx. Yield 50—6-ounce servings	
Ingredients	**Amount**	**Market Price**	**Extension Cost**
Beef Chuck	18 lb.	$1.35 per lb.	$24.30
Garlic, Minced	1 oz.	1.52 per lb.	.095
Flour	8 oz.	.14 per lb.	.07
Chili Powder	3/4 oz.	1.76 per lb.	.08
Paprika	5 oz.	2.38 per lb.	.74
Tomato Puree	1 qt.	1.20 per gal.	.30
Water	8 lb.	—	—
Bay Leaves	2	—	—
Caraway Seeds	1/2 oz.	2.75 per lb.	.085
Onions, Minced	2 lb.	.15 per lb.	.30
Salt	1 oz.	.32 per lb.	.02
Pepper	1/4 oz.	1.28 per lb.	.02
		Total Cost	$26.010
		Cost Per Serving	.52

Fig. 16-2 Standardized recipe.

ACHIEVEMENT REVIEW

COSTING STANDARD RECIPES

Complete the following cost charts, showing the extension cost, the total cost, and the cost per serving. Carry the price three places to the right of the decimal.

1.

Braised Swiss Steak		Approx. Yield 50 servings	
Ingredients	**Amount**	**Market Price**	**Extension Cost**
6 oz. Round Steaks	50 ea.	$1.45 per lb.	
Onions	12 oz.	.18 per lb.	
Garlic	1 oz.	1.52 per lb.	
Tomato Puree	1 pt.	1.20 per gal.	
Brown Stock	6 qt.	.90 per gal.	
Salad Oil	3 cups	.70 per qt.	
Bread Flour	12 oz.	.14 per lb.	
Salt	3/4 oz.	.16 per lb.	
		Total Cost	
		Cost Per Serving	

2.

Salisbury Steak		Approx. Yield 50 servings	
Ingredients	**Amount**	**Market Price**	**Extension Cost**
Beef Chuck	14 lb.	$1.15 per lb.	
Onions	3 lb.	.15 per lb.	
Garlic	1/2 oz.	1.52 per lb.	
Salad Oil	1/2 cup	.70 per qt.	
Bread Cubes	2 lb.	.36 per lb.	
Milk	1 1/2 pt.	1.65 per gal.	
Whole Eggs	8	.84 per doz.	
Pepper	1/4 oz.	1.18 per lb.	
Salt	1 oz.	.16 per lb.	
		Total Cost	
		Cost Per Serving	

3.

Buttermilk Biscuits		Approx. Yield 6 dozen = 72 biscuits	
Ingredients	**Amount**	**Market Price**	**Extension Cost**
Cake Flour	1 lb. 8 oz.	$ .18 per lb.	
Bread Flour	1 lb. 8 oz.	.14 per lb.	
Baking Powder	3 1/2 oz.	.75 per lb.	
Salt	1/2 oz.	.16 per lb.	
Sugar	4 oz.	.13 per lb.	
Butter	12 oz.	.80 per lb.	
Buttermilk	2 lb. 4 oz.	.36 per qt.	
		Total Cost	
		Cost Per Dozen	
		Cost Per Biscuit	

4.

Brown Sugar Cookies		Approx. Yield 14 dozen = 168 cookies	
Ingredients	**Amount**	**Market Price**	**Extension Cost**
Brown Sugar	3 lb. 2 oz.	$ .28 per lb.	
Shortening	2 lb. 4 oz.	.34 per lb.	
Salt	1 oz.	.16 per lb.	
Baking Soda	1/2 oz.	.38 per lb.	
Pastry Flour	4 lb. 8 oz.	.16 per lb.	
Whole Eggs	9	.84 per doz.	
Vanilla	1/4 oz.	.78 per pt.	
		Total Cost	
		Cost Per Dozen	
		Cost Per Cookie	

5.

Soft Dinner Rolls		Approx. Yield 16 dozen = 192 rolls	
Ingredients	**Amount**	**Market Price**	**Extension Cost**
Sugar	1 lb.	$ .13 per lb.	
Shortening	1 lb. 4 oz.	.34 per lb.	
Dry Milk	8 oz.	.24 per lb.	
Salt	2 oz.	.16 per lb.	
Whole Eggs	3	.84 per doz.	
Yeast	6 oz.	.56 per lb.	
Water	4 lb.	—	
Bread Flour	7 lb.	.14 per lb.	
		Total Cost	
		Cost Per Dozen	
		Cost Per Roll	

6.

Stuffed Spare Ribs		Approx. Yield 50 servings	
Ingredients	**Amount**	**Market Price**	**Extension Cost**
Spare Ribs	30 lb.	$1.56 per lb.	
Pork Picnic, Fresh	4 lb.	.86 per lb.	
Bread, Fresh or Dried	2 lb.	.36 per lb.	
Milk	1 1/2 qt.	1.65 per gal.	
Onions	8 oz.	.15 per lb.	
Celery	8 oz.	.39 per lb.	
Margarine	8 oz.	.45 per lb.	
Parsley	1 bch.	.20 per bch.	
Sage	1/4 oz.	.95 per lb.	
Salt	1/4 oz.	.16 per lb.	
Pepper	1/8 oz.	1.28 per lb.	
Egg Yolks	4	.84 per doz.	
		Total Cost	
		Cost Per Serving	

7.

White Cake		Approx. Yield 14—8-inch cakes	
Ingredients	**Amount**	**Market Price**	**Extension Cost**
Cake Flour	2 lb. 8 oz.	$ .18 per lb.	
Shortening	1 lb. 12 oz.	.34 per lb.	
Granulated Sugar	3 lb. 2 oz.	.13 per lb.	
Salt	1 1/2 oz.	.16 per lb.	
Baking Powder	2 1/2 oz.	.75 per lb.	
Water	14 oz.	—	
Dry Milk	2 1/2 oz.	.24 per lb.	
Whole Eggs	6	.84 per doz.	
Egg Whites	1 lb.	1.28 per lb.	
Water	1 lb.	—	
Vanilla	1/8 oz.	.96 per pt.	
		Total Cost	
		Cost Per Cake	

8.

Apple Pie Filling		Approx. Yield 6—8-inch pies	
Ingredients	**Amount**	**Market Price**	**Extension Cost**
Apples	1 #10 can	$1.78 per #10 can	
Apple Juice	1 lb. 8 oz.	.76 per qt.	
Granulated Sugar	1 lb. 4 oz.	.13 per lb.	
Salt	1/4 oz.	.16 per lb.	
Cinnamon	1/4 oz.	1.88 per lb.	
Nutmeg	1/8 oz.	1.68 per lb.	
Butter	3 oz.	.80 per lb.	
Water	8 oz.	—	
Modified Starch	3 oz.	.38 per lb.	
		Total Cost	
		Cost Per Pie	

9.

Soft Rye Rolls		Approx. Yield 12 dozen rolls	
Ingredients	**Amount**	**Market Price**	**Extension Cost**
Bread Flour	6 lb. 6 oz.	$ .14 per lb.	
Dark Rye Flour	1 lb. 4 oz.	.15 per lb.	
Yeast	6 oz.	.56 per lb.	
Salt	1 3/4 oz.	.16 per lb.	
Dry Milk	5 oz.	.24 per lb.	
Shortening	1 lb.	.34 per lb.	
Granulated Sugar	1 lb.	.13 per lb.	
Malt	1 1/2 oz.	.89 per lb.	
Water	4 lb. 8 oz.	—	
Caraway Seeds	6 oz.	1.24 per lb.	
		Total Cost	
		Cost Per Dozen	

10.

Deviled Crabs		Approx. Yield 25 Servings	
Ingredients	**Amount**	**Market Price**	**Extension Cost**
King Crab Meat	6 lb.	$3.65 per lb.	
Shortening	12 oz.	.34 per lb.	
Flour	12 oz.	.14 per lb.	
Onions	12 oz.	.18 per lb.	
Milk	1 1/2 qt.	1.65 per gal.	
Prepared Mustard	1/4 cup	.42 per pt.	
Worcestershire Sauce	1 oz.	1.18 per pt.	
Tabasco Sauce	1 tsp.	—	
Lemon Juice	1/4 cup	.69 per qt.	
Sherry Wine	1 cup	2.94 per gal.	
Salt	1/4 oz.	.16 per lb.	
Pepper	1/8 oz.	1.28 per lb.	
		Total Cost	
		Cost Per Serving	

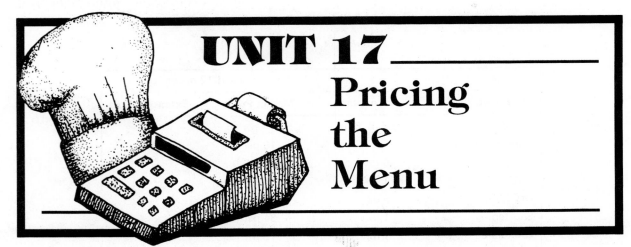

UNIT 17

Pricing the Menu

There is no one standard method of pricing a menu. Many things must be considered before a menu price is determined. The cost of all items purchased, rent, labor cost, equipment, taxes, and so forth, are usually considered before deciding how much must be charged to make a profit.

Larger operations usually have accountants and computers to supply an accurate picture of the overall cost, which makes the pricing decision much easier. In smaller operations, the overall cost is more difficult to compute because controls are not always as tight as they should be and records are not always accurate because of the time and cost involved.

MARKUP

In menu pricing, the only standard seems to be that cost must be established before a *markup* (amount added to raw food cost to obtain a selling price) can be added to determine a menu price. The amount of markup usually varies depending on the type of establishment. A cafeteria might mark up all of its items by one-half the cost to obtain a menu price; a gourmet restaurant might mark up all its items by two or three times the cost to obtain a menu price. The markup is not always figured using fractions. In many cases, percents are used because they are easier to work with and less mistakes are made.

Example: Amount of markup using a fraction. If the raw food cost is $.95 and the markup rate is ⅔, multiply the cost by the markup rate.

$$\frac{.95}{1} \times \frac{2}{3} = \frac{1.90}{3} = .633$$

Add the markup (.633) to the raw food cost to determine the selling cost or menu price.

$$\begin{array}{r} \$\ .95 \\ +\ \ .633 \\ \hline \$1.583 \text{ or } \$1.58 \end{array}$$

If the markup is figured by using percent, multiply the raw food cost by the percent and add the markup to the raw food cost to determine the menu or selling price. When multiplying or dividing with percents, it is best to convert the percent to its decimal equivalent. This is done by removing the percent sign (%) and moving the decimal point two places to the left.

Examples:

$$60\% = .60$$
$$75\% = .75$$
$$25.5\% = .255$$
$$50.5\% = .505$$

Example: Amount of markup using a percent.

If the raw food cost is $.95 and the markup rate is 45 percent, convert 45 percent to .45 and multiply.

$$
\begin{array}{r}
\$ \ .95 \\
\times \ \ .45 \\
\hline
475 \\
380 \ \ \\
\hline
\$.4275 = \$.43
\end{array}
$$

Add the markup ($.43) to the raw food cost to determine the selling cost or menu price.

$$
\begin{array}{r}
\$ \ .95 \\
+ \ .43 \\
\hline
\$1.38
\end{array}
$$

FOOD COST PERCENT

Another method of menu pricing is to determine the monthly food cost percent and divide the food cost percent into the raw food cost.

Example: If the raw food cost is $.75 and the monthly food cost percent is 40 percent, the price is determined by dividing 40 percent (or .40) into $.75.

$$
\begin{array}{r}
\$ \ 1.875 = \$1.88 \text{ selling price} \\
.40 \,)\overline{\$.75.000} \\
\underline{40} \\
35\ 0 \\
\underline{32\ 0} \\
3\ 00 \\
\underline{2\ 80} \\
200 \\
\underline{200}
\end{array}
$$

This method is sometimes used in a simpler manner by determining that to maintain a 33 percent food cost, the selling or menu price of a meal or item should be 3 times the raw food cost. (For 40 percent food cost, the menu price should be 2½ times the raw food cost; for a 50 percent food cost, the menu price should be 2 times the raw food cost.)

Example: If the raw cost is $.75, and the monthly food cost percent is 40 (2½ times the raw food cost), the price is determined by multiplying 2½ by $.75.

$$2\frac{1}{2} \times \$.75 = \frac{5}{2} \times \$.75 = \frac{\$3.75}{2}$$
$$= \$1.875 \text{ or } \$1.88 \text{ selling price}$$

This problem can also be done by first converting 2½ to a decimal before multiplying:

$$
\begin{array}{r}
\$ \ .75 \\
\times \ \ 2.5 \\
\hline
375 \\
150 \ \ \\
\hline
\$1.875 = \$1.88 \text{ selling price}
\end{array}
$$

It has become a custom in most food service establishments to price menu items so that they end in amounts of $.25, $.50, $.75, or $1.00 to speed the totaling of checks. The customer can relate to the price faster, and in most cases less change is handled. This helps the establishment to provide good service and maintain a more efficient operation. Thus, when a menu or selling price is determined to be $2.18, the price listed on the menu is usually $2.25.

Examples:

Determined Price	Menu Price
$1.15	$1.25
$2.35	$2.50
$3.60	$3.75
$4.90	$5.00

Some restaurants have been successful in listing menu prices that end in odd cents, such as $.69, $1.39, $2.49, etc. This, however, is not the usual practice.

Remember that prices are important in a customer's appraisal of a food service operation. Try to establish a price level that appeals to all types of potential customers. This does not only mean a price they can afford, but one that they feel is closely related to the quality of food and service they receive.

Also remember that the operator is in business to make a profit. Profits will usually continue if an establishment is built on a solid foundation of quality food, fair prices, and good service.

ACHIEVEMENT REVIEW
PRICING THE MENU

Determine the menu price if:

1. The raw food cost is $.98 and the markup rate is ¾.

2. The raw food cost is $1.58 and the markup rate is ⅔.

3. The raw food cost is $2.69 and the markup rate is ⅞.

4. The raw food cost is $1.67 and the markup rate is ⅝.

5. The raw food cost is $1.44 and the markup rate is ⅗.

6. The raw food cost is $.99 and the markup rate is 75%.

7. The raw food cost is $1.86 and the markup rate is 66%.

8. The raw food cost is $3.18 and the markup rate is 68%.

9. The raw food cost is $1.35 and the markup rate is 42.5%.

10. The raw food cost is $2.55 and the markup rate is 70%.

11. The raw food cost is $1.76 and the markup rate is 45.5%.

12. The raw food cost is $1.38 and the markup rate is 66.6%.

13. The raw food cost is $2.05 and the markup rate is 48%.

14. The raw food cost is $2.38 and the markup rate is 33.3%.

15. The raw food cost is $3.15 and the markup rate is 35.5%.

16. The raw food cost is $1.65 and the food cost percent for the month is 45%.

17. The raw food cost is $2.18 and the food cost percent for the month is 38%.

18. The raw food cost is $3.28 and the food cost percent for the month is 37%.

19. The raw food cost is $2.75 and the food cost percent for the month is 35%.

20. The raw food cost is $1.98 and the food cost percent for the month is 43%.

21. The raw food cost is $2.90 and the food cost percent for the month is 52%.

22. The raw food cost is $1.39 and the food cost percent for the month is 40%.

23. The raw food cost is $3.48 and the food cost percent for the month is 33%.

24. The raw food cost is $1.25 and the food cost percent for the month is 35%.

25. The raw food cost is $1.73 and the food cost percent for the month is 45%.

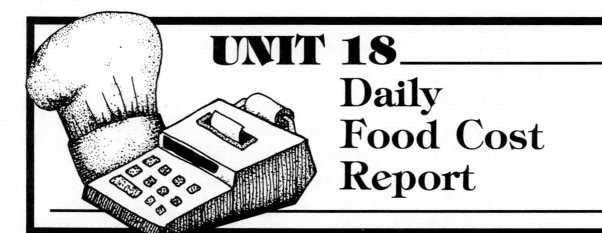

UNIT 18

Daily Food Cost Report

The *daily food cost report* is a tool that is used by management to control the use and cost of food. Its purpose is to show the exact cost and amount of food used on any given day. This report is a guide that keeps the manager aware of costs, thus helping to control the cost of food being used in the establishment. With the high cost of food and much waste on the part of employees, it is essential that tight controls be maintained on the items that can lead a food service operation to bankruptcy.

The daily food cost report also helps to provide a more accurate picture of the monthly food cost pattern. That is, if the planned monthly food cost percent is set at a certain rate to ensure a profitable operation and each day this rate is exceeded, something is not right. It would therefore be apparent to management that all factors such as portion size, waste, and food production must be investigated to find the cause.

There are several kinds of food cost reports in use, but the simplest type is made up from the totals of storeroom requisitions and direct purchases for the day. The example shown in figure 18-1 is a simplified form of the daily food cost report.

A *storeroom requisition* is a list of food items issued from the storeroom upon the request of the production crew. These may be requests for certain meats, other groceries, canned foods, or frozen foods. The requisitions are priced and extended at the end of each day. The unit price is usually marked on all storeroom items as they are received and stored so the requisitioned foods can be priced immediately. *Direct purchases* are those foods that are usually purchased each day or every other day. They include produce, dairy products, fresh seafood, bread, rolls, and any other items that are considered perishable. The daily purchase total of direct purchases can be obtained from the *invoice* (a list of goods sent to a purchaser with their prices and quantity listed) which is sent with each order.

In the example in figure 18-1, the total sale for the day is $360.00. This figure is found by totaling the register sales for the day. The figure can be checked by multiplying the average sales by the customer count.

$$
\begin{array}{rl}
\$ \quad 1.20 & \text{Average Sales} \\
\times \ 300 & \text{Customer Count} \\
\hline
\$360.00 & \text{Total Sales}
\end{array}
$$

			Last Month
Issues	**Today**	**To Date**	**To Date**
Storeroom			
Canned Goods	$ 25.50	$ 100.50	$ 95.25
Other Groceries	15.60	45.25	43.50
Meat	35.25	225.50	222.30
Frozen Foods	10.60	30.40	25.50
Direct Purchases			
Poultry, fresh	20.15	75.30	74.35
Seafood, fresh	21.30	58.60	57.60
Produce	8.00	39.25	35.20
Dairy Products	6.00	25.00	20.70
Bread and Rolls	5.00	20.35	18.40
Miscellaneous	4.00	10.50	10.45
Total Cost	$151.40	$ 630.65	$ 603.25
Total Sales	$360.00	$1,500.00	$1,400.00
Food Cost Percent	42.1%	42%	43%

Customer Count <u>300</u>
Average Sale <u>$1.20</u>

Date <u>January 15, 19</u>
Day <u>Thursday</u>

Fig. 18-1 Daily food cost report.

The customer count is obtained by checking the register which records the total number of customers as sales are rung up. The amount of average sales for the day is found by dividing the customer count into the total sales:

$$\begin{array}{r} \text{Customer} \quad \$\ \ 1.20 \text{ Average Sales} \\ \text{Count} \quad 300\overline{)\$360.00} \text{ Total Sales} \\ \underline{300} \\ 600 \\ \underline{600} \end{array}$$

The cost of food for the day is $151.40. This figure is found by adding together the cost of all items issued for the day's food production. The food cost percent is 42.1 percent. This is considered a good food cost percent in most food service establishments since most owners and managers strive to keep this rate under 45 percent. To find this ratio in terms of percent, the cost of food ($151.40) is divided by the total sales for the day ($360.00).

$$\begin{array}{r} .421 = 42.1\% \text{ Food percent} \\ \$360\overline{)\$151.400} \quad \text{Cost of food} \end{array}$$

In addition to the daily food cost percent, some daily food cost reports also show the increasing totals for the month and the totals of the previous month to date. (See figure 18-1.) By including this information, the owner or manager has a clear picture of the food cost pattern.

185

ACHIEVEMENT REVIEW

DAILY FOOD COST REPORT

For the following daily food cost reports find:

- Total cost of food for today, to date, and last month to date.
- Total sales for today.
- Food cost percent today, to date, and last month to date.

1.

Customer Count 390 Average Sale $1.45		Date July 5, 19 Day Monday	
Issues	**Today**	**To Date**	**Last Month To Date**
Storeroom			
Canned Goods	$24.30	$ 106.55	$ 98.40
Other Groceries	17.80	40.50	42.20
Meat	47.25	325.50	312.30
Frozen Foods	9.50	25.22	23.60
Direct Purchases			
Poultry, Fresh	18.15	75.45	72.53
Seafood, Fresh	20.35	52.60	48.22
Produce	10.56	37.65	34.25
Dairy Products	6.90	23.75	19.85
Bread and Rolls	5.85	24.43	23.46
Miscellaneous	6.20	12.24	11.47
Total Cost	—	—	—
Total Sales	—	$1,650.00	$1,560.00
Food Cost Percent	—	—	—

2.

Customer Count 290		Date August 3, 19	
Average Sale $1.50		Day Tuesday	

Issues	Today	To Date	Last Month To Date
Storeroom			
Canned Goods	$30.24	$ 110.50	$ 112.25
Other Groceries	15.20	40.65	45.50
Meat	58.65	228.60	248.40
Frozen Foods	10.90	30.75	25.95
Direct Purchases			
Poultry, Fresh	17.55	68.30	75.35
Seafood, Fresh	25.40	59.42	65.75
Produce	15.50	49.25	52.20
Dairy Products	8.22	22.00	20.80
Bread and Rolls	8.45	18.80	21.50
Miscellaneous	6.30	12.43	13.65
Total Cost	—	—	—
Total Sales	—	$1,600.00	$1,700.00
Food Cost Percent	—	—	—

3.

Customer Count <u>380</u> Average <u>$1.35</u>		Date <u>September 8, 19</u> Day <u>Wednesday</u>	
Issues	**Today**	**To Date**	**Last Month To Date**
Storeroom			
Canned Goods	$36.75	$ 115.25	$ 116.35
Other Groceries	16.26	42.75	43.85
Meat	68.40	245.30	240.75
Frozen Foods	12.65	28.80	27.60
Direct Purchases			
Poultry, Fresh	18.95	58.35	85.62
Seafood, Fresh	24.15	46.12	58.14
Produce	18.85	39.25	48.20
Dairy Products	9.95	20.90	21.76
Bread and Rolls	10.22	18.80	22.87
Miscellaneous	7.50	9.20	13.70
Total Cost	—	—	—
Total Sales	—	$1,550.00	$1,590.00
Food Cost Percent	—	—	—

4.

| Customer Count 365 | | Date October 7, 19 | |
| Average Sale $2.20 | | Day Thursday | |

Issues	Today	To Date	Last Month To Date
Storeroom			
Canned Goods	$45.60	$ 125.25	$ 119.25
Other Groceries	25.55	63.23	62.15
Meat	75.95	324.34	305.60
Frozen Foods	15.40	40.36	35.75
Direct Purchases			
Poultry, Fresh	22.25	68.90	62.42
Seafood, Fresh	34.65	49.50	48.60
Produce	19.20	42.20	39.95
Dairy Products	11.55	29.60	24.33
Bread and Rolls	16.21	23.30	21.15
Miscellaneous	10.64	15.65	14.60
Total Cost	—	—	—
Total Sales	—	$1,825.00	$1,760.00
Food Cost Percent	—	—	—

5.

Customer Count 295 Average Sale $3.25		Date November 5, 19 Day Friday	
Issues	**Today**	**To Date**	**Last Month To Date**
Storeroom			
Canned Goods	$54.90	$ 134.22	$ 136.25
Other Groceries	38.70	68.95	69.90
Meat	95.75	395.80	398.95
Frozen Foods	30.20	50.60	54.90
Direct Purchases			
Poultry, Fresh	32.45	78.91	80.96
Seafood, Fresh	44.67	62.40	67.50
Produce	23.83	46.92	48.22
Dairy Products	14.94	36.25	39.81
Bread and Rolls	18.22	32.95	33.96
Miscellaneous	13.82	19.65	21.85
Total Cost	—	—	—
Total Sales	—	$2,250.00	$2,165.00
Food Cost Percent	—	—	—

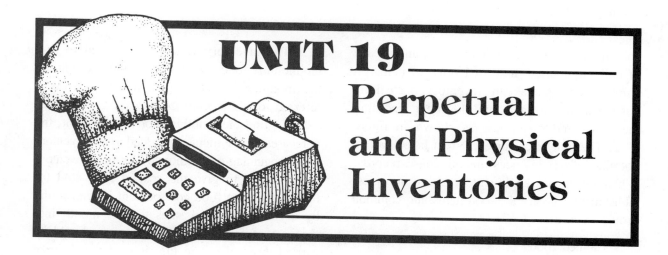

UNIT 19
Perpetual and Physical Inventories

PERPETUAL INVENTORY

A *perpetual inventory* is a continuous or endless inventory. It is a record which is taken in the storeroom to show the balance on hand for each storeroom item, figure 19-1. As a requisition from the production crew is received, and the items are issued, the amounts are subtracted from the inventory balance. When new shipments of each item arrive and are placed on the shelves, they are added to the inventory balance. At the end of each month, when the physical inventory of each item is taken, the physical inventory should match the perpetual inventory. If the two do not match, a problem exists. The problem may be poor bookkeeping, negligence in checking incoming orders, theft, or a similar situation.

PHYSICAL INVENTORY

The *physical inventory* is taken at the end of each month in order to determine the accurate cost of food consumed during that period. When doing a physical inventory, an actual count is taken of all stock on hand. The physical inventory figures are most important when making out the profit and loss statement.

The inventory sheet should be prepared in advance, using a standard form. The form should

Fig. 19-1 In most food service operations, a storeroom inventory is taken at the end of each month.

contain the name of each item, the quantity, size or bulk, unit price, and total price of items on hand. Two people take the inventory, one calling out the items on hand, and the other recording the items and quantities, as shown in figure 19-2. The unit price for each item listed can be obtrained from invoices and purchase records. When setting up an inventory sheet, most establishments classify the food items into common groups such as:

- Canned foods
- Other groceries
- Butter, eggs, and cheese
- Coffee and tea
- Fruits and vegetables
- Meats, poultry, and fish
- Supplies

When the total inventory value is found, this figure represents the cost of food still on hand or in storage. It is food that has not been sold during this period.

Fig. 19-2 Students taking a storeroom inventory.

The inventory value is then used to find the monthly food cost percent. When computing the monthly food cost percent, this figure is referred to as the *final inventory*.

To find the monthly food cost percent, the purchases for the month are added to the inventory at the beginning of the month. The final inventory is subtracted to give the cost of food sold for the month. The food cost percent is then found by dividing the total sales for the month into the cost of food sold. The division should be carried three places to the right of the decimal to give the exact percent. See figure 19-3.

The amount of sales is found by totaling the register tapes for that particular month. It is the total of all the sales made during that period. The purchases for the month represent the total of all the food purchased during that month. This is found by adding the totals of all invoices sent with each food purchase. The inventory at the beginning of the month is the final inventory from the previous month. (An example of a physical inventory is given on pages 194 through 200.)

Example: The final inventory for the month of October is the beginning inventory for the month of November. The final inventory, as stated before, is found from the physical inventory taken at the end of each month.

The monthly food cost percent tells the restaurant operator what percentage of the total sales for that period was used to purchase food for the operation. This is a very important figure and one that must be controlled, because the cost of food and labor are the highest costs in any food service operation.

Sales $2,550.00

Inventory at Beginning of Month   300.00

Purchases for the Month   895.00

Final Inventory   250.00

To find the *Cost of Food Sold*, first add:

Inventory at Beginning of Month $ 300.00
to Purchases for the Month   + 895.00
     $1,195.00

Then, from this sum, subtract the Final   – 250.00
Inventory $ 945.00 (Cost of Food Sold)

To find the *Monthly Food Cost Percent*, divide the

Sales into the Cost of Food Sold:

$$\begin{array}{r} .370 \\ 2{,}550 \overline{)945.000} \\ \underline{765\ 0} \\ 180\ 00 \\ \underline{178\ 50} \\ 1\ 500 \end{array}$$

= 37% (Monthly Food Cost Percent)

Fig. 19-3 How to find cost of food sold and monthly food cost percent.

Example of a Physical Inventory

(Prices are not Current)

Weekly or Period

Inventory Recapitulation

Week Ending

August 9, 19 ____

Period Ending

August 31, 19 __

Item No.	Item	Amount
1.	Canned Goods	$ 215.37
2.	Other Groceries	218.13
3.	Butter, Eggs, and Cheese	34.31
4.	Coffee and Tea	20.07
5.	Fruits and Vegetables	111.97
6.	Meat, Poultry, and Fish	464.23
	Total Food	$1,064.08
7.	Supplies	213.63
	Total Inventory Value	$1,277.71

Called by *Joe Jones*

Extended by *Ron Sharp*

Manager *John Sparks*

Canned Goods			
Quantity and Size	**Item**	**Unit Price**	**Extension**
8-#10	Apples	$ 1.22	$ 9.76
6-#10	Apricots	1.43	8.58
9-#10	Beans, Green	1.18	10.62
10-#10	Beans, Kidney	.98	9.80
12-#10	Beans, Wax	1.15	13.80
14-#10	Bean Sprouts	.89	12.46
4-#10	Beets, Whole	1.12	4.48
3-#10	Carrots, Whole	1.14	3.42
8-#10	Cherries	1.30	10.40
2-#10	Fruit Cocktail	1.28	2.56
3-#10	Noodles (Chow Mein)	.79	2.37
6-#10	Peach Halves	1.32	7.92
5-#10	Peaches, Pie	1.25	6.25
5-#10	Pears	1.36	6.80
4-#10	Pineapple Tid-bits	1.30	5.20
10-#10	Pineapple Slices	1.38	13.80
4-#10	Plums	1.35	5.40
6-#10	Pumpkin	1.06	6.36
7-#10	Sweet Potatoes	1.02	7.14
9-#10	Tomato Catsup	1.20	10.80
8-#10	Tomato Puree	1.23	9.84
3-#10	Tomatoes	1.25	3.75
15-#2	Asparagus Spears	.58	8.70
18-#2	Cream Style Corn	.40	7.20
20-#2	Whole Kernel Corn	.45	9.00
24-#2	Salmon	.79	18.96
		Canned Goods Total	$215.37

Groceries — Dry Bulk Goods			
Quantity and Size	**Item**	**Unit Price**	**Extension**
8 lb.	Baking Soda	$ .16	$ 1.28
9 lb.	Baking Powder, 5 lb. box	2.75	4.95
8 lb.	Cocoa, 5 lb. box	3.45	5.52
6 lb.	Coconut Shred	.46	2.76
10 lb.	Cracker Meal	.26	2.60
4 lb.	Chicken Base	1.06	4.24
7 lb.	Beef Base	1.06	7.42
6 lb.	Raisins	.48	2.88
9 lb.	Cornmeal	.22	1.98
12 lb.	Cornstarch	.23	2.76
20 lb.	Tapioca Flour	.28	5.60
63 lb.	Bread Flour, 100 lb. bag	9.90	6.24
53 lb.	Cake Flour, 100 lb. bag	12.50	6.63
74 lb.	Pastry Flour, 100 lb. bag	10.80	7.99
15 lb.	Elbow Macaroni	.59	8.85
18 lb.	Spaghetti	.20	3.60
6 lb.	Rice	.22	1.32
12 lb.	Noodles	.18	2.16
		Subtotal	$ 78.78

Groceries — Oils and Fats			
Quantity and Size	**Item**	**Unit Price**	**Extension**
12 lb.	Margarine, 32 lb. case	$ 6.72	$ 2.52
42 lb.	Shortening, 50 lb. tin	15.00	12.60
3 1/2 gal.	Salad Oil, 5 gal. can	6.40	4.48
		Subtotal	$ 19.60

Groceries — Spices			
Quantity and Size	**Item**	**Unit Price**	**Extension**
1 1/2 lb.	Allspice, Ground	$ .92	$ 1.38
1 3/4 lb.	Bay Leaves	.88	1.54
2 lb.	Chili Powder	.69	1.38
3 lb.	Cinnamon	1.12	3.36
4 lb.	Cloves, Ground	.89	3.56
1 1/4 lb.	Ginger	.92	1.15
1 1/2 lb.	Comino Seed	.94	1.41
2 1/2 lb.	Celery Seed	.85	2.13
2 1/4 lb.	Mustard Seed	.92	2.07
1 3/4 lb.	Mace	.96	1.68
2 3/4 lb.	Mustard, Dry	.94	2.59
1 1/2 lb.	Marjoram	.86	1.29
3 lb.	Nutmeg	.84	2.52
4 lb.	Oregano	.93	3.72
5 lb.	Pepper, White	1.08	5.40
3 lb.	Pepper, Black	1.02	3.06
2 lb.	Pickling Spices	.68	1.36
1 1/2 lb.	Rosemary Leaves	.78	1.17
1 1/4 lb.	Sage	.56	.70
1 3/4 lb.	Thyme	.68	1.19
		Subtotal	$ 42.66

Groceries — Dressing and Condiments			
Quantity and Size	**Item**	**Unit Price**	**Extension**
3 gal.	Mayonnaise	$ 3.75	$ 11.25
2 gal.	Dill Pickles	1.45	2.90
1 1/2 gal.	Sweet Relish	1.65	2.48
4 gal.	Salad Dressing	2.40	9.60
6 1/2 gal.	Vinegar	1.33	8.65
1 1/4 gal.	Choy Sauce	1.10	1.38
4 gal.	Worcestershire Sauce	2.26	9.04
5 gal.	French Dressing	2.84	14.20
		Subtotal	$ 59.50

Groceries — Coloring and Extracts			
Quantity and Size	**Item**	**Unit Price**	**Extension**
1 1/2 qt.	Caramel Color	$ 1.10	$ 1.65
1 1/4 pt.	Yellow	1.24	1.55
1 3/4 pt.	Red	1.32	2.31
2 pt.	Green	1.33	2.66
1 pt.	Lemon Extract	1.65	1.65
3 pt.	Vanilla Extract	1.75	5.25
1 1/2 pt.	Maple Extract	1.68	2.52
		Subtotal	$ 17.59
		Other Groceries Total	$218.13

Butter, Eggs, and Cheese			
Quantity and Size	**Item**	**Unit Price**	**Extension**
8 lb.	American Cheese	$ 1.15	$ 9.20
15 lb.	Chip Butter	.89	13.35
14 doz.	Eggs	.84	11.76
		Butter, Eggs, and Cheese Total	$ 34.31

Coffee and Tea			
Quantity and Size	**Item**	**Unit Price**	**Extension**
16 lb.	Coffee	$ 1.08	$ 17.28
1/2 pkg.	Tea—indiv. (100)	2.98	1.49
1/4 pkg.	Tea—ice (48)	5.20	1.30
		Coffee and Tea Total	$ 20.07

Fruits and Vegetables — Fresh			
Quantity and Size	Item	Unit Price	Extension
4 bu.	Carrots	$ .20	$ .80
3 lb.	Endive	.78	2.34
7 head	Head Lettuce	.39	2.73
8 lb.	Leaf Lettuce	.78	6.24
20 lb.	Dry Onions	.16	3.20
54 lb.	Red Potatoes	.12	6.48
44 lb.	Idaho Potatoes	.18	7.92
6 lb.	Tomatoes	.58	3.48
3 lb.	Parsley	.20	.60
5 lb.	Green Peppers	.28	1.40
8 lb.	Apples	.16	1.28
7 lb.	Bananas	.18	1.26
4 doz.	Lemons	.79	3.16
5 doz.	Oranges	.89	4.45
5 bu.	Radishes	.15	.75
3 bu.	Celery	.33	.99
		Subtotal	$ 47.08

Fruits and Vegetables — Frozen			
Quantity and Size	Item	Unit Price	Extension
9 lb.	Strawberries	$ .38	$ 3.42
12 lb.	Peaches	.42	5.04
16 lb.	Blueberries	.48	7.68
34 lb.	Lima Beans	.32	10.88
25 lb.	Corn	.31	7.75
18 lb.	Broccoli	.36	6.48
12 lb.	Brussels Sprouts	.38	4.56
18 lb.	Mixed Vegetables	.28	5.04
22 lb.	Peas	.34	7.48
16 lb.	Cauliflower	.41	6.56
		Subtotal	$ 64.89
		Fruits and Vegetables Total	$111.97

Meat, Poultry, and Fish			
Quantity and Size	Item	Unit Price	Extension
15 lb.	Beef Ground	$ 1.10	$ 16.50
21 lb.	Beef Round	1.49	31.29
30 lb.	Beef Ribs	2.21	66.30
22 lb.	Beef Rib Eye	2.45	53.90
19 lb.	Beef Chuck	1.15	21.85
15 lb.	Club Steak	2.45	36.75
14 lb.	Beef Tenderloin	3.10	43.40
28 lb.	Pork Loin	.98	27.44
22 lb.	Boston Butt	.95	20.90
36 lb.	Veal Leg	1.84	66.24
12 lb.	Veal Shoulder	1.23	14.76
14 lb.	Veal Loin	1.55	21.70
32 lb.	Ham	1.35	43.20
		Meat, Poultry, and Fish Total	$464.23

Supplies			
Quantity and Size	Item	Unit Price	Extension
68 T	Napkins	$ 1.65	$112.20
8 T	Butter Chips	1.59	12.72
22 T	Soufflé Cups	.98	21.56
15 T	Paper Bags	2.79	41.85
12 gal.	Bleach	.39	4.68
13 gal.	Dish	1.03	13.39
21 lb.	Salute	.18	3.78
5 ea.	Pot Brushes	.69	3.45
		Supplies Total	$213.63

Sales	$16,290.00
Inventory at Beginning of Month	2,180.50
Purchases for the Month	4,880.00
Final Inventory	1,277.71
Cost of Food Sold	5,782.79
Food Cost Percent	35.4%

ACHIEVEMENT REVIEW
INVENTORIES

A. In problems 1 through 10, find the cost of food sold and the monthly food cost percent. Carry percents three places to the right of the decimal.

1. Sales $2,000.00
   - Inventory at Beginning of Month   150.00
   - Purchases for the Month   650.00
   - Final Inventory   200.00
   - Cost of Food Sold   Find
   - Monthly Food Cost Percent   Find

2. Sales $5,000.00
   - Inventory at Beginning of Month   250.00
   - Purchases for the Month   3,000.00
   - Final Inventory   450.00
   - Cost of Food Sold   Find
   - Monthly Food Cost Percent   Find

3. Sales $4,250.00
   - Inventory at Beginning of Month   250.00
   - Purchases for the Month   2,175.00
   - Final Inventory   350.00
   - Cost of Food Sold   Find
   - Monthly Food Cost Percent   Find

4. Sales $2,550.00
   - Inventory at Beginning of Month   300.00
   - Purchases for the Month   895.00
   - Final Inventory   250.00
   - Cost of Food Sold   Find
   - Monthly Food Cost Percent   Find

5. Sales $4,675.00
   - Inventory at Beginning of Month   450.00
   - Purchases for the Month   1,650.00
   - Final Inventory   325.00
   - Cost of Food Sold   Find
   - Monthly Food Cost Percent   Find

6. Sales $6,725.00
   - Inventory at Beginning of Month   700.00
   - Purchases for the Month   3,785.00
   - Final Inventory   550.00
   - Cost of Food Sold   Find
   - Monthly Food Cost Percent   Find

7. Sales $1,575.50
   Inventory at Beginning of Month   385.50
   Purchases for the Month   890.25
   Final Inventory   450.20
   Cost of Food Sold   Find
   Monthly Food Cost Percent   Find

8. Sales $2,380.00
   Inventory at Beginning of Month   365.00
   Purchases for the Month   958.00
   Final Inventory   268.00
   Cost of Food Sold   Find
   Monthly Food Cost Percent   Find

9. Sales $10,780.00
   Inventory at Beginning of Month   1,985.00
   Purchases for the Month   7,250.00
   Final Inventory   1,650.00
   Cost of Food Sold   Find
   Monthly Food Cost Percent   Find

10. Sales $3,760.00
    Inventory at Beginning of Month   495.00
    Purchases for the Month   685.00
    Final Inventory   290.00
    Cost of Food Sold   Find
    Monthly Food Cost Percent   Find

B. In this exercise, the teacher calls out the quantity and unit price of each item listed. Quantity and unit price may be true and current value or hypothetical. The instructor should also give the figures of sales, inventory at beginning of month, and purchases for the month.

The student is asked to find the extension price, subtotals, totals, total food (on front of inventory), total inventory value (on front of inventory), and the monthly food cost percent (using the procedure shown in figure 19-3).

Form to Use for Part B. Weekly or Period Inventory Recapitulation		
Week Ending		Period Ending
Item No.	**Item**	**Amount**
1.	Canned Goods	
2.	Other Groceries	
3.	Butter, Eggs, and Cheese	
4.	Coffee and Tea	
5.	Fruits and Vegetables	
6.	Meat, Poultry, and Fish	
	Total Food	
7.	Supplies	
	Total Inventory Value	
Called by _____		Extended by _____ Manager _____

Canned Goods			
Quantity and Size	**Item**	**Unit Price**	**Extension**
#10	Apples		
#10	Apricots		
#10	Beans, Green		
#10	Beans, Kidney		
#10	Beans, Wax		
#10	Bean Sprouts		
#10	Beets, Whole		
#10	Carrots, Whole		
#10	Cherries		
#10	Fruit Cocktail		
#10	Noodles (Chow Mein)		
#10	Peach Halves		
#10	Peaches, Pie		
#10	Pears		
#10	Pineapple Tid-bits		
#10	Pineapple Slices		
#10	Plums		
#10	Pumpkin		
#10	Sweet Potatoes		
#10	Tomato Catsup		
#10	Tomato Puree		
#10	Tomatoes		
# 2	Asparagus Spears		
#22	Cream Style Corn		
# 2	Whole Kernel Corn		
# 2	Salmon		
Canned Goods Total			

Groceries — Dry Bulk Goods			
Quantity and Size	**Item**	**Unit Price**	**Extension**
lb.	Baking Soda		
lb.	Baking Powder		
lb.	Cocoa		
lb.	Coconut Shred		
lb.	Cracker Meal		
lb.	Chicken Base		
lb.	Beef Base		
lb.	Raisins		
lb.	Cornmeal		
lb.	Cornstarch		
lb.	Tapioca Flour		
lb.	Bread Flour		
lb.	Cake Flour		
lb.	Pastry Flour		
lb.	Elbow Macaroni		
lb.	Spaghetti		
lb.	Rice		
lb.	Noodles		
Subtotal			

Groceries — Oils and Fats			
Quantity and Size	**Item**	**Unit Price**	**Extension**
lb.	Margarine		
lb.	Shortening		
lb.	Salad Oil		
Subtotal			

Groceries — Spices			
Quantity and Size	**Item**	**Unit Price**	**Extension**
lb.	Allspice, Ground		
lb.	Bay Leaves		
lb.	Chili Powder		
lb.	Cinnamon		
lb.	Cloves, Ground		
lb.	Ginger		
lb.	Comino Seed		
lb.	Celery Seed		
lb.	Caraway Seed		
lb.	Mace		
lb.	Mustard, Dry		
lb.	Marjoram		
lb.	Nutmeg		
lb.	Oregano		
lb.	Pepper, White		
lb.	Pepper, Black		
lb.	Pickling Spices		
lb.	Rosemary Leaves		
lb.	Sage		
lb.	Thyme		
	Subtotal		

Groceries — Dressing and Condiments			
Quantity and Size	**Item**	**Unit Price**	**Extension**
gal.	Mayonnaise		
gal.	Dill Pickles		
gal.	Sweet Relish		
gal.	Salad Dressing		
gal.	Vinegar		
gal.	Choy Sauce		
gal.	Worcestershire Sauce		
gal.	French Dressing		
	Subtotal		

Groceries — Coloring and Extracts			
Quantity and Size	**Item**	**Unit Price**	**Extension**
qt.	Caramel Color		
pt.	Yellow		
pt.	Red		
pt.	Green		
pt.	Lemon Extract		
pt.	Vanilla Extract		
pt.	Maple Extract		
Subtotal			
Other Groceries Total			

Butter, Eggs, and Cheese			
Quantity and Size	**Item**	**Unit Price**	**Extension**
lb.	American Cheese		
lb.	Chip Butter		
doz.	Eggs		
Butter, Eggs, and Cheese Total			

Coffee and Tea			
Quantity and Size	**Item**	**Unit Price**	**Extension**
lb.	Coffee		
pkg.	Tea—indiv. (100)		
pkg.	Tea—ice (48)		
Coffee and Tea Total			

Fruits and Vegetables — Fresh			
Quantity and Size	**Item**	**Unit Price**	**Extension**
bu.	Carrots		
lb.	Endive		
head	Head Lettuce		
lb.	Leaf Lettuce		
lb.	Dry Onions		
lb.	Red Potatoes		
lb.	Idaho Potatoes		
lb.	Tomatoes		
bu.	Parsley		
lb.	Green Peppers		
lb.	Apples		
lb.	Bananas		
doz.	Lemons		
doz.	Oranges		
bu.	Radishes		
bu.	Celery		
Subtotal			

Fruits and Vegetables — Frozen			
Quantity and Size	**Item**	**Unit Price**	**Extension**
t	Strawberries		
lb.	Peaches		
lb.	Blueberries		
lb.	Lima Beans		
lb.	Corn		
lb.	Broccoli		
lb.	Brussels Sprouts		
lb.	Mixed Vegetables		
lb.	Peas		
lb.	Cauliflower		
Subtotal			
Fruits and Vegetables Total			

Meat, Poultry, and Fish			
Quantity and Size	**Item**	**Unit Price**	**Extension**
lb.	Beef Ground		
lb.	Beef Round		
lb.	Beef Ribs		
lb.	Beef Rib Eyes		
lb.	Beef Chuck		
lb.	Club Steak		
lb.	Beef Tenderloin		
lb.	Pork Loin		
lb.	Boston Butt		
lb.	Veal Leg		
lb.	Veal Shoulder		
lb.	Veal Loin		
lb.	Ham		
	Subtotal		
	Meat, Poultry, and Fish Total		

Supplies			
Quantity and Size	**Item**	**Unit Price**	**Extension**
T	Napkins		
T	Butter Chips		
T	Soufflé Cups		
T	Paper Bags		
gal.	Bleach		
gal.	Dish		
lb.	Salute		
ea.	Pot Brushes		
	Supplies Total		

Sales
Inventory at Beginning of Month
Purchases for the Month
Final Inventory
Cost of Food Sold
Food Cost Percent

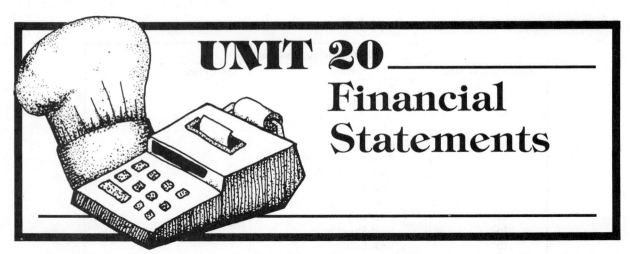

UNIT 20
Financial Statements

In any business there are two major financial statements which must be prepared and are essential to the operation of that business. These financial statements are especially important to today's restaurant operator. Prices fluctuate so rapidly it is necessary to be aware of the business' financial position at all times. These two financial statements are the profit and loss (income) statement and balance sheet.

PROFIT AND LOSS STATEMENT

The *profit and loss statement* is a summary or report of the business operation for a given period of time. The purpose of the statement is to determine how much money the business is making or losing. In the profit and loss statement all income from sales is set off against expenses to determine the profit or loss. The following formula summarizes the profit and loss statement of a restaurant operation.

Sales minus Cost of Food Sold =
Gross Margin
Gross Margin minus Total Operation
Expenses = *Net Profit or Loss*

The net profit or loss is the figure that is of greater concern to the restaurant operator because it determines the success or failure of the business.

Most of the figures used in making a profit and loss statement are taken from the daily records which are recorded and made available by the bookkeeper, figure 20-1. The daily records of the restaurant operation are kept in a book called the *cashbook*. The cashbook contains a record of all income and expenses. It is a means of

Fig. 20-1 Bookkeeper working on one of the financial statements required in a food service operation.

keeping track of every sales dollar. The cashbook is also important for securing the figures needed for tax purposes and other financial obligations.

A profit and loss statement can be made up whenever the restaurant operator wants to know the business' financial situation. For example, it can be done every month, every three months, every six months, or even once a year. When the statement is completed, the restaurant operator analyzes all the figures and compares the dollar amounts in each category with those of previous months or previous years. The operator may also wish to make a percent comparison of all figures with total sales, which represents 100 percent. The example of a profit and loss statement, figure 20-2, shows dollar amounts and percentages of the total sales. Percents are usually the language spoken by the restaurant operator. For example, when speaking of food cost or labor cost, a percentage of total sales is always given. The operator may state "My food cost was 35 percent last month" or "My labor cost was 60 percent last month," etc. Percents, rather than dollar amounts, give a clearer picture of the overall operation. It is said that if food cost and labor cost exceed 75 percent of the total sales, the operation is in trouble.

Listings on the Profit and Loss Statement

Total sales. Found by totaling the register tapes for that particular month. This amount is also found in the cashbook.

Inventory at beginning of month. The final inventory from the previous month. Example: the final inventory for the month of October is the beginning inventory for the month of November.

Purchases for the month. The total of all food purchased during the month. This is also found in the cashbook.

Total. The sum of the inventory at the beginning of the month and the purchases for the month. It is a total of all food that was available during that month.

Final inventory. Acquired from the physical inventory taken at the end of each month's operation. It represents the cost of food that is still in stock and was not sold. Since it represents food that was not sold, it is subtracted from the total cost of food that was available during the month.

Cost of food sold. When the final inventory is subtracted from the cost of food on hand, the result is the cost of food sold that month.

Gross margin (gross profit). Found by subtracting the cost of food sold from the total sales. *Gross* means the total value before deductions are made. *Margin* is the difference between the cost and the selling price, so sales minus cost of food gives gross margin.

Salaries (labor cost). The labor cost consists of all wages paid to all employees, including the owner's salary. Except for the cost of food, this is the most important item to control.

Social security taxes. These taxes are paid to the government for the purpose of retirement. The employer must deduct a certain percent of the employee's salary for social security taxes and at the same time, must equal the amount

Profit and Loss Statement

For the Month of September, 19___

Total Sales	$12,000	100.0%
Food Cost:		
Inventory at Beginning of Month	$1,210	
Purchases for the Month	4,700	
Total	5,910	
Less: Final Inventory	750	
Cost of Food Sold	$ 5,160	43.0%
Gross Margin	$ 6,840	57.0%
Expenses		
Salaries	$3,840	32.0%
Social Security Taxes	240	2.0%
Rent	600	5.0%
Laundry and Linens	84	.7%
Repairs and Maintenance	336	2.8%
Advertising	60	.5%
Taxes and Insurance	180	1.5%
Supplies	84	.7%
Depreciation	240	2.0%
Utilities	360	3.0%
Miscellaneous Expenses	180	1.5%
Total Operating Expenses	$ 6,204	51.7%
Net Profit	$ 636	5.3%

Fig. 20-2 Profit and loss statement.

the employee pays. The amount shown on the profit and loss statement represents only the amount the employer must pay.

Rent. A fixed amount paid at a certain time of each month to the owner of a property for the use of that property. If a restaurant operator owns the building in which the restaurant is located, this particular expense is not incurred. However, there may be a similar expense if the building is not paid for (for example, paying off the mortgage to a building and loan company or a bank). A restaurant operator who rents usually has a long term lease with the owner of the property, so the operator can be assured of continued occupancy for a specific period of time.

Laundry and linens.   These items include the cost of cleaning or renting all uniforms, napkins, tablecloths, towels, and so forth.

Repairs and maintenance.   Expenses that result from equipment failure or building repairs if the restaurant property is owned. If the property is rented, the owner may assume the responsibility for all repairs. This item is an essential part of any restaurant business because equipment must be kept in good shape at all times for an efficient operation.

Advertising.   An expense that is brought about when notifying the public about a place of business. The advertising may be done through the newspaper, television, radio, billboards, periodicals, or other media.

Taxes and insurance.   Usually paid on an annual basis. If these are listed on the monthly statement, as done in figure 20-2, the monthly cost is found by taking one-twelfth of the yearly payment.

Supplies   The cost of those items used other than food. These include janitor supplies, paper products, and similar expenses.

Depreciation.   The act of lessening the value of an item as it wears out. For example, if a new mixing machine is bought for $900.00 and is used constantly for one year, at the end of

that year it is worth less than $900.00. It has been worn out slightly through use, and therefore, has less value. There are several methods used for figuring depreciation but the simplest and most practical is the straightline method.

In the straightline method, estimate the length of time a piece of equipment is expected to last, and the trade-in value it should possess at the end of its estimated life. The difference between the original cost of the equipment and its estimated trade-in value gives the total amount of allowable depreciation. For example, the mixing machine that was purchased for $900.00 is expected to last 12 years, at which time it will probably have a trade-in value of $60.00.

Original cost	$900.00
Estimated trade-in	60.00
Allowable depreciation	$840.00

The allowable depreciation ($840.00) is divided by the number of years the item is expected to last (12).

$$
\begin{array}{r}
\$\ 70.00 \\
12\overline{)\$840.00} \\
\underline{84} \\
0 \\
\underline{0}
\end{array}
$$

$ 70.00 Amount that can be deducted each of the 12 years

Some restaurant operators estimate that all their major pieces of equipment must be replaced every 10 or 12 years, so instead of figuring depreciation on individual pieces of equipment they figure it on the group. Thus each year they charge off one-tenth or one-twelfth of the allowable depreciation. If a depreciation figure for a month is needed, find the depreciation for one year and take one-twelfth of that amount.

Utilities. The cost incurred through the supply of gas, electricity, and water. These bills are usually based on monthly use and are presented to the customer on a monthly basis.

Miscellaneous expenses. Usually smaller than the others listed, they include licenses, organization dues, charitable contributions, and so forth. These amounts vary with the size and policy of the operation.

ACHIEVEMENT REVIEW
PROFIT AND LOSS STATEMENT

Prepare profit and loss statements using the amounts given in the problems listed below. Use the same form as shown in the example in figure 20-2. Find the cost of food sold, gross margin, total operating expenses, net profit, and the percent of sales for each problem listed.

1. The Manor Restaurant had total sales for the month of November of $15,500. Their inventory at the beginning of the month was $5,280. During the month they made purchases that totaled $8,200. The final inventory at the end of the month was $4,690.

   The Manor Restaurant had the following expenses during the month: salaries $3,220, social security taxes $115, rent $460, laundry $95, repairs and maintenance $421, advertising $75, taxes and insurance $195, supplies $120, depreciation $540, utilities $380, and miscellaneous expenses $210.

2. Connie's Cafeteria had total sales for the month of March of $25,830. Their inventory at the beginning of the month was $7,275. During the month they made purchases that totaled $10,900. The final inventory at the end of the month was $6,870.

   Connie's Cafeteria had the following expenses during the month: salaries $5,225, social security taxes $313.50, rent $540, laundry and linens $98, repairs and maintenance $268, advertising $78, taxes and insurance $218, supplies $120, depreciation $395, utilities $270, and miscellaneous expenses $168.

3. The Golden Goose Restaurant had total sales for the month of April of $75,300. Their inventory at the beginning of the month was $8,800. During the month they made purchases that totaled $25,500. The final inventory at the end of the month was $8,440.

   The Golden Goose Restaurant had the following expenses during the month: salaries $14,350, social security taxes $875, rent $650, laundry and linens $225, repairs and maintenance $563, advertising $156, taxes and insurance $419, supplies $280, depreciation $690, utilities $345, and miscellaneous expenses $242.

4. The Branding Iron Steak House had total sales for the month of September of $30,522. Their inventory at the beginning of the month was $5,220. During the month they made purchases that totaled $10,425. The final inventory at the end of the month was $5,835.

    The Branding Iron Steak House had the following expenses: salaries $6,225, social security taxes $421, rent $525, laundry and linens $105, repairs and maintenance $376, advertising $85, taxes and insurance $258, supplies $148, depreciation $528, utilities $285, and miscellaneous expenses $188.

5. Bob's Drive-in had total sales for the month of May of $18,258. Their inventory at the beginning of the month was $4,780. During the month they made purchases that totaled $7,660. The final inventory at the end of the month was $5,225

    Bob's Drive-in had the following expenses: salaries $3,120, social security taxes $187, rent $275, laundry and linens $87, repairs and maintenance $192, advertising $78, taxes and insurance $156, supplies $93, depreciation $432, utilities $204, and miscellaneous expenses $112.

THE BALANCE SHEET

The *balance sheet* is a statement listing the company's *assets* (what is owned) and *liabilities* (what is owed) to determine net worth, proprietorship, or capital. It is a dollar and cents picture of their financial status at a given time. The given time is usually December 31 if based on the calendar year, or June 30 if based on the fiscal year. A balance sheet can be prepared by a company at any time they wish to know their net worth; however, in most cases this occurs only once a year. The balance sheet is, of course, a necessary part of business operations, but it can also be used by individuals. It serves many purposes: to provide necessary information in reporting financial matters to the state and federal government for income taxes, to secure loans from banks or building and loan companies, and in the case of many business operations, to provide the information stockholders and partners want to see in the annual report.

The formula for preparing a balance sheet can be expressed by the following equation:

Assets − Liabilities = Net Worth (or Proprietorship)

In other words, a company totals all the money it owes and subtracts that amount from the total value of all it owns to find out its total worth.

To prove that the balance sheet is correct, the liabilities are added to the net worth. The sum should equal the total assets. The relationship may be expressed by this equation:

Assets = Liabilities + Proprietorship

Two examples of balance sheets are shown in figure 20-3, page 217. One is for an individual, the other is for a restaurant operation.

Listings on the Balance Sheet

The amounts shown on the William Jones balance sheet come from various sources, as indicated below.

Home. What a home is worth on the current market can be determined by a real estate adjuster or, in some cases, a salesperson.

Home furnishings. This amount is more difficult to figure. Probably the best method is to list the original cost and deduct a certain percentage for wear.

Savings certificates. These are easy to total since the amounts are stated on each certificate.

Cash in a savings account. This is stated in the passbook (the last balance entered in the book).

Automobile, houseboat. The current worth of an automobile or houseboat may be found by checking with the company from which they were purchased.

The figures shown on the balance sheet for the Charcoal King Restaurant come from many sources, but can be obtained from records kept on file by the restaurant, and in particular, the bookkeeper.

Cash. The cash amount is found in the cash-

Example A

William Jones
Balance Sheet, December 31, 19___

Assets		Liabilities	
Home	$55,000	Home Mortgage	$22,000
Home furnishings	6,490	Houseboat loan	4,500
Savings certificates	10,000	Auto loan	1,500
(2) $5,000 ea.			
Cash in savings accounts	4,500	Charge accounts (retail stores)	690
Cash in checking account	295	Total Liabilities	$28,690
Automobile	4,650	Net Worth—Proprietorship	$61,020
Houseboat	8,775	Total Liabilities and	
Total Assets	$89,710	Net Worth—Proprietorship	$89,710

Example B

Charcoal King Restaurant
Balance Sheet, June 30, 19___

Assets (Current)		Liabilities (Current)	
Cash (in bank and on hand)	$ 5,450	Accounts payable	$ 3,679
Accounts receivable	1,675	Installment accounts or	
Food and beverage inventory	2,585	bank notes payable	12,156
Supplies	420	Payroll and occupational	
Total Current Assets	$10,130	taxes payable	989
		Total Current Liabilities	$16,824
Assets (Fixed)		Net Worth—Proprietorship	$12,256
Stationary kitchen equipment	$10,322	Total Liabilities and	
Hand kitchen equipment	2,456	Net Worth—Proprietorship	$29,080
Dining room furniture and			
fixtures	3,387		
China, glassware, silver,			
and linen	2,785		
Total Fixed Assets	$18,950		
Total Assets (Current and			
Fixed	$29,080		

Fig. 20-3 (A) Balance sheet for an individual. (B) Balance sheet for a company.

book. The cashbook is kept up to date by the bookkeeper.

Accounts receivable. This refers to money that is owed to the bookkeeper.

Food and beverage inventory. This is taken from the profit and loss statement.

Assets (fixed). The value of the fixed assets listed (which, in this case, are for various kinds of equipment), can be acquired by referring to purchase contracts or equipment records.

Accounts payable. This refers to money the restaurant owes for purchases made (in other words, bills that have not been paid). The amounts may be found by referring to the unpaid bill or invoice file.

Installment accounts or bank notes payable. These are found by referring to sales contracts or equipment records. Another possible source is the bank or company to whom the money is owed.

Payroll and occupational taxes payable. These are found by checking with the bookkeeper or accountant who keeps these figures up to date.

ACHIEVEMENT REVIEW
BALANCE SHEET

Prepare balance sheets for each of the following, using the figures listed. Find total assets, total liabilities, net worth, and total liabilities and net worth—proprietorship.

1. Mr. John Brown had the following assets as of June 20, this year: home $35,000, savings certificates (4) @ $5,000 each, cash in savings account $4,800, cash in checking account $695, and automobile $2,850. His liabilities were as follows: home mortgage $18,700, auto loan $1,500, and charge accounts $962.

2. Mr. Tim Woodard had the following assets as of May 15, this year: home $46,000, home furnishings $8,695, automobile $4,250, speedboat $2,150, cash in savings account $12,890, cash in checking account $1,678, and stocks $1,275. His liabilities were as follows: home mortgage $28,500, auto loan $2,400, boat loan $1,300, note payable to loan company $1,500, and charge accounts $894.

3. Joe and Mary Jones had the following assets as of July 15, this year: home $52,500, house trailer $3,700, home furnishings $7,775, automobile $3,860, cash in savings account $6,265, cash in checking account $253, U.S. Savings Bonds $850, and savings certificates (2) @ $5,000 each. Their liabilities were as follows: home mortgage $28,500, loan on house trailer $2,200, automobile loan $2,300, and charge accounts $897.

4. Mr. R.C. Stone had the following assets as of July 20, this year: home $43,500, home furnishings $8,880, two automobiles (one $5,440, the other $3,820), cash in savings account $2,435, cash in checking account $925,

U.S. Savings Bonds $1,250, and a $5,000 savings certificate. His liabilities were as follows: home mortgage $31,500, loan on first automobile $3,550, loan on second automobile $1,800, and charge accounts $980.

5. Robert and Dolores Hunt had the following assets as of December 31, this year: local home $38,750, vacation home $15,280, motorboat $2,800, automobile $6,550, home furnishings $5,990, savings certificates (4) @ $5,000 each, cash in savings account $1,200, cash in checking account $565, and U.S. Savings Bonds $425. Their liabilities were as follows: local home mortgage $22,900, vacation home mortgage $10,550, motorboat loan $1,300, automobile loan $3,450, loan on home furnishings $1,800, and charge accounts $900.

6. The Seaside Restaurant had the following assets as of June 30, this year: cash $3,289, accounts receivable $1,325, food and beverage inventory $2,500, supplies $436.50, stationary kitchen equipment $6,290, hand equipment $595, dining room furniture and fixtures $4,400, and china, glassware, silver, and linen $2,125. The restaurant's liabilities were as follows: accounts payable $6,595, installment accounts or bank notes payable $8,290, and payroll and occupational taxes payable $855.

7. Jim's Hamburger Palace had the following assets as of December 31, this year: cash $1,489, food and beverage inventory $1,200, supplies $98, equipment (stationary, hand, and serving equipment) $3,535, and furniture and fixtures $1,250. The liabilities were as follows: accounts payable $86.20, note payable to First National Bank $975, note payable to Mike Bryce $280, sales tax payable $69.50, and occupational tax payable $260.

8. Junior's Drive-In had the following assets as of December 31, this year: cash $3,165.75, accounts receivable $290, food and beverage inventory $3,090, supplies $550, stationary and hand equipment $7,280, serving equipment and dishes $2,900, and furniture and fixtures $2,695. The liabilities were as follows: accounts payable $586, note payable to Second National Bank $2,860, social security tax $286, and state sales tax $298.

9. Bill's Oyster House had the following assets as of December 31, this year: cash $4,692.50, accounts receivable $546.58, food and beverage inventory $1,495.50, supplies $389.90, stationary kitchen equipment $5,449.35, hand equipment $428.55, dining room furniture and fixtures $3,338.50, and serving equipment and dishes $2,876.75. The liabilities were as follows:

accounts payable $6,297.80, installment accounts payable $2,670.50, bank notes payable to First National Bank $7,525, sales tax payable $265.40, payroll tax payable $280.50, and occupational tax payable $60.95.

10. The Colony Restaurant had the following assets as of June 30, this year: cash $5,658.25, accounts receivable $626.48, food and beverage inventory $2,995.25, supplies $735.45, all kitchen equipment $4,868.50, dining room furniture and fixtures $5,600.50, and serving equipment $1,970.79. The liabilities were as follows: accounts payable $6,950.25, installment accounts payable $1,975.75, bank note payable to First National Bank $2,525, sales tax payable $138.40, payroll tax payable $159.60, and occupational tax payable $75.75.

UNIT 21
Break-even Analysis

Break-even analysis was introduced to restaurant operators by William P. Fisher, Ph.D. through a pamphlet titled "Profitable Financial Management for Food Service Operators." The pamphlet is published by the National Restaurant Association and is available to all members.

Break-even analysis is a method of calculating that level of economic activity where a business neither makes a profit nor incurs a loss. It is based on a certain figure that must be maintained throughout total sales before a profit can be realized and below which a loss is incurred. This method is beneficial to the restaurant operator for planning profits. Planning profits is very important to a business operation when financial planning decisions must be made.

The concept of determining the break-even point of a restaurant operation can be expressed in this formula:

> Sales Revenue – Variable Cost = Gross Profit
> Gross Profit – Fixed Cost = 0

Example:

Sales Revenue – Variable Cost = Gross Profit   Gross Profit – Fixed Cost = 0

$10,000 – $5,000 = $5,000

$ 5,000 – $5,000 = 0 Break-even Point

Mathematical problems are expressed by formulas. If the formula is followed, a solution is found. It is important to be able to identify and be familiar with each part of the break-even formula.

Sales Revenue. The money received from the sale of all products, food, beverage, etc. It represents all of the money received over a certain period of time through cash sales.

Variable Cost. Changeable cost. Expenses that change with the level of sales volume, that is, the greater the number of people served and meals produced the greater the variable cost. Variable cost may be divided into two categories: cost of goods sold and serving expenses.

Cost of goods sold is the amount it costs the operator to provide food, beverages, etc. to serve customers. It is the operator's cost of goods to be sold.

Serving expenses are additional expenses incurred as a result of serving customers. A few are listed:

- laundry—tablecloths, napkins, kitchen towels, uniforms

- paper supplies—guest checks, report forms, register tapes
- part-time payroll—additional help on busy days
- tableware replacement–glassware, china, silverware

Gross Profit.   Sales revenue minus variable costs. The total value before deducting fixed cost.

Fixed Cost.   Expenses that remain constant regardless of the level of sales volume. They may be divided into two categories: occupational expenses and primary expenses. Occupational expenses are those expenses which can be called the cost of ownership, and continue whether or not the restaurant is operating. They include such expenses as property tax, insurance, interest on mortgage, and depreciation on equipment, furniture, and building. There may be others depending on the local situation.

Primary expenses are those which result from being open for business. These expenses are constant whether serving a few customers or many. They include such expenses as utilities, basic staff (preparation, service, etc.), telephone service, repairs and maintenance, licenses, and exterminating cost.

In the example, there is a $10,000 sales revenue and $5,000 variable expenses. The operation would therefore have a $5,000 gross profit. If the fixed cost is $5,000, this restaurant would have zero profit and zero loss. It thus breaks even. The $10,000 sales level is the break-even point for this business. Anything over this figure is profit; anything below is a loss.

The relationship between sales revenue, variable costs, and fixed costs must be known for profit planning purposes. It is necessary to find the percents that can be applied to these different dollar amounts.

Taking the figures given in the example and putting them into the form of an operating statement, the percents are as follows:

Sales Revenue	$10,000	100%
Less Variable Costs	5,000	50%
= Gross Profit	5,000	50%
Less Fixed Cost	5,000	50%
= Net Income (loss)	–0– (break-even)	

Remember that percents are used by the restaurant operator because they give a clearer picture of the overall relationship between sales revenue and cost. The percents shown in the example were found in the following way:

Sales revenue always equals 100 percent. It represents all that is received. It was shown earlier in this text that to find a percent, the whole is divided into the part, so in this case $10,000 is divided into the variable cost $5,000.

$$\frac{\$5,000}{\$10,000} = \begin{array}{l} 50\% \text{ and } 10,000 \text{ into gross} \\ \text{profit } \$5,000 = 50\% \end{array}$$

Variable costs plus gross profit = 100 percent. Next, $10,000 is divided into the fixed cost $5,000.

$$\frac{5,000}{10,000} = 50 \text{ percent fixed cost}$$

Then 50 percent is subtracted from the gross profit, to get 0 percent (break-even).

In calculating the break-even point, two key facts should be remembered:

- Sales revenue always equals 100 percent.
- Once the gross profit percent is determined, it is easy to calculate the break-even point.

In this example, sales equals $10,000, which represents 100 percent. The gross profit is 50 percent which is determined by dividing the dollar amount of gross profit by sales.

$$(\frac{5,000}{10,000} = .50 \text{ gross profit percent})$$

Having found the gross profit percent, it is divided into the fixed cost to determine the break-even point.

$\underline{\$5,000}$ Fixed cost
.50 Gross profit percent

$$\begin{array}{r} 10,000.00 \\ .50\overline{)\ 5,000.00} \\ \underline{5\ 0} \\ 0 \end{array} = \begin{array}{l} \text{Break-even point} \\ \text{sales volume} \end{array}$$

Problem: The variable cost of the Blue Jay Restaurant is 40 percent of the total sales. The fixed cost is $18,000. What is the break-even point?

Solution: Sales   100%
       Variable Cost   40% Given
       Gross Profit Percent   60% Found

Divide 60 percent (gross profit percent) into $18,000 (fixed cost) to find the break-even point.

$$\begin{array}{r} \$30,000. \\ .60\overline{)\$18,000.00} \\ \underline{18\ 0} \\ 0 \end{array} \text{break-even point}$$

Proof:

Sales	$30,000	100%	Always
Variable Cost	12,000	40%	Known
Gross Profit	18,000	60%	Found
Fixed Cost (Known)	$\underline{18,000}$		
Net Income (Loss)	–0–		Break-even

Of course, breaking even is not the reason a person is in business. One is usually in business to make a profit. Assume that the Blue Jay Restaurant did $50,000 in sales and determine the profit through break-even analysis.

It is known that 40 percent of every dollar taken in is used to pay the cost of food and beverage. It is also known that the fixed cost is $18,000. So knowing these figures it is simple to determine what the profit will be by following the formula given previously.

Sales Revenue	$50,000	100%	Always
Less Variable Cost	20,000	40%	Known
Gross Profit	30,000	60%	Found
Less Fixed Cost (Known)	18,000		
Net Income—Before Taxes	$12,000	Profit	

In this calculation, it is noted that in dollar amounts, variable cost is $20,000. The variable cost rate of 40 percent remains the same. Remember that the percent remains constant but the dollar amounts fluctuate with an increase or decrease in sales volume. This is because it costs more money (in food, beverage, laundry, tableware–all variable costs) to serve more people, and less money to serve less people.

The break-even point indicates that at a given sales volume the fixed cost has been met. Fixed means that a certain portion of cost does not change regardless of the volume of business.

Looking at the problem again, the following rule can be proven: all sales above the break-even point result in a profit (before taxes) of an amount equal to the gross profit percent.

Proof:

$50,000	Sales Revenue
− $30,000	Break-even point
$20,000	Above break-even point (profit)

Gross profit rate (60%) times $20,000

$20,000
× .60
$12,000 Yes, this is equal to the profit before taxes.

The rule holds true. It always will.

There are times in the restaurant business when a loss occurs.

Example: Assume that the Blue Jay Restaurant did only $28,000 in sales.

Sales Revenue	$28,000	100% Always
Variable Cost	11,200	40% Known
Gross Profit	16,800	60% Found
Fixed Cost (Known)	18,000	
Net Loss	$ 1,200	

Note that although the variable cost percent rate remained constant (40%), the dollar amount changed.

$28,000 Sales
× .40 Variable Cost
$11,200 Dollar amount of variable cost

The change occurred because of a decrease in sales volume from $50,000 (when a profit was made) to $28,000 (when a loss occurred).

ACHIEVEMENT REVIEW
BREAK-EVEN ANALYSIS

For each problem given below, determine the break-even point, show proof, and find the amount of profit or loss.

If a profit exists, see if the rule about gross profit percent holds true.

1. The Blue Bird Cafeteria did $80,000 in sales. Their variable cost was 40 percent of the total sales and the fixed costs were $24,000.

2. The Castle Restaurant did $120,000 in sales. Their variable cost was 36 percent of the total sales and the fixed costs were $32,000.

3. The Blue Angel Restaurant did $32,000 in sales. Their variable cost was 40 percent of the total sales and the fixed costs were $24,000.

4. The Red Gate Cafeteria did $96,000 in sales. Their variable cost was 43 percent of the total sales and the fixed costs were $63,600.

5. The Dutch Mill Restaurant did $83,894 in sales. Their variable cost was 43 percent of the total sales and the fixed costs were $63,600.

6. The Yellow Rose Restaurant did $75,000 in sales. Their variable cost was 42 percent of the total sales and the fixed costs were $23,200.

7. The Old Mill Restaurant did $72,500 in sales. Their variable cost was 44 percent of the total sales and the fixed costs were $28,112.

8. The White Pine Restaurant did $40,100 in sales. Their variable cost was 45 percent of the total sales and the fixed costs were $22,165.

9. Jim and Joe's Cafeteria did $61,600 in sales. Their variable cost was 32 percent of the total sales and the fixed costs were $34,408.

10. Bob's Drive-In Restaurant did $29,400 in sales. Their variable cost was 36 percent of the total sales and the fixed costs were $19,200.

UNIT 22
Budgeting

A budget is an estimate of probable income and expenditures for a calendar or fiscal year. A budget helps individuals, small businesses, or corporations maintain a fairly equal balance between income and expenses.

Not all food service operations find it necessary or advantageous to budget. Those that do are usually large corporations that operate their own food service for the benefit of their employees. Their goal for the operation is usually to break even or to hold their losses to a minimum.

In a budget the areas listed under the two major components, income and expenditures, will vary, depending on the size of the overall operation. Generally included under income are the total of register receipts and income from sales that would not pass through the register, such as payment for a catered affair. These sales might be labeled supplemental income. These figures can be obtained from income records of the previous year.

Expenditures might include:

- Total payroll or labor cost
- Total supplies (food, paper goods, cleaning supplies, etc.)
- Telephone service
- New equipment
- Service calls on equipment
- Workmen's compensation
- Utilities (gas and electric)
- Printing service for menus
- Fringe benefits (dental plan, hospitalization, etc.)

Many of the estimated figures required for these items can again be obtained from the records of the previous years. If after checking the figures from the previous year, you are aware of or believe that an increase or decrease of funds will be required for the coming year then adjustments are certainly in order. Remember, the purpose of a budget is to give a general idea of what will happen. The budget is very seldom exact.

Using hypothetical figures, let us set up a budget as if it were a pie. The whole pie would be the food service operation's total income for a year. Two examples will be given in the form of problems. Example 1 will show a budget in percent and ask you to find the dollar amounts. Example 2 will show dollar amounts and ask you to find the percent.

Example 1: A food service budget (break-even budget) is shown in figure 22-1. The yearly in-

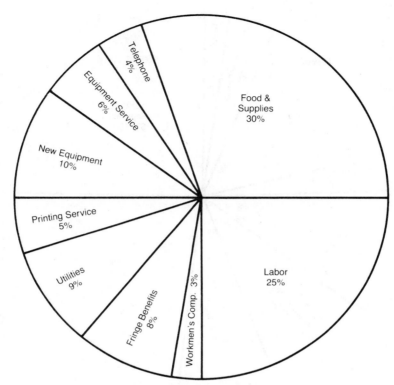

Fig. 22-1 Sample budget

come from register and supplemental sales is $300,000. Find the total dollar amounts that were budgeted for each item listed.

Food and Supplies (30%)	=	$90,000
Labor (25%)	=	$75,000
Workmen's Compensation (3%)	=	$ 9,000
Fringe Benefits (8%)	=	$24,000
Utilities (9%)	=	$27,000
Printing Service (5%)	=	$15,000
New Equipment (10%)	=	$30,000
Equipment Service (6%)	=	$18,000
Telephone (4%)	=	$12,000

When taking a percent of a whole number, in this case $300,000, we must multiply. For example,

```
   300,000    yearly income
×     .30    food & supplies
$90,000.00    amount budgeted for food
                & supplies
```

```
 $300,000    yearly income
×     .25    labor
  1500000
  600000
$75,000.00    amount budgeted for labor
```

The other amounts shown in the budget are found by the same procedure.

Example 2: A food service budget (break-even budget) is shown in figure 22-2. The yearly income from register and supplemental sales is

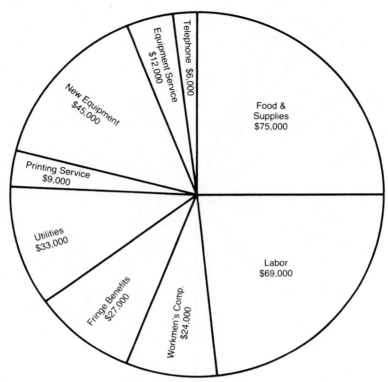

Fig. 22-2 Sample budget

$300,000. Find the percentage of the total income that was budgeted for each item listed. The sum of the completed percentages should equal 100%.

Food and Supplies ($75,000)	=	25%
Labor ($69,000)	=	23%
Workmen's Compensation ($24,000)	=	8%
Fringe Benefits ($27,000)	=	9%
Utilities ($33,000)	=	11%
Printing Service ($9,000)	=	3%
New Equipment ($45,000)	=	15%
Equipment Service ($12,000)	=	4%
Telephone ($6,000)	=	2%

To find what percent one number is of another number, divide the number that represents the part by the number that represents the whole. For example,

$$\begin{array}{r} .25 = 25\% \\ \$300,000\overline{)\$75,000.00} \\ \underline{600000} \\ 1500000 \\ \underline{1500000} \end{array}$$

food & supplies represents the part

Thus, 25% is the percentage of the $300,000 yearly income set aside for the cost of food and supplies. The other percentages shown in the budget are found by the same procedure.

Budgeting is also an important step in the

proper management of personal affairs. The family income certainly should be budgeted, so the money will be on hand when the bills are due. The family budget may be set up on a monthly or yearly basis, with a percentage of income or a dollar amount set aside for each important family expense.

Example: For a monthly income of $2000.00 the budget might be as follows:

Item	Percent	Dollar Amount
Food	25%	$500.00
Clothing	15%	$300.00
Mortgage Payment	35%	$700.00
Recreation	9%	$180.00
Benevolences	5%	$100.00
Savings	11%	$220.00
	100%	$2000.00

ACHIEVEMENT REVIEW
THE BUDGET

1. For a food service budget the yearly income from register and supplemental sales is $550,000.00. Find the dollar amount that was budgeted for each item listed.

1.	Food and supplies	26%	$ _____
2.	Labor	24%	_____
3.	Workman's Compensation	4%	_____
4.	Fringe Benefits	8%	_____
5.	Utilities	11%	_____
6.	Printing Service	2%	_____
7.	New Equipment	10%	_____
8.	Equipment Service	5%	_____
9.	Telephone	3%	_____
10.	Profit	7%	_____

2. For a food service budget the yearly income from register and supplemental sales is $670,000.00. Find the dollar amount that was budgeted for each item listed.

1.	Food and Supplies	26%	$ _____
2.	Labor	22%	_____
3.	Workman's Compensation	4%	_____
4.	Fringe Benefits	9%	_____
5.	Utilities	11%	_____
6.	Printing Service	7%	_____

7.	New Equipment	8%	_____
8.	Equipment Service	6%	_____
9.	Telephone	2%	_____
10.	Profit	5%	_____

3. For a food service budget the yearly income from register and supplemental sales is $725,000.00. Find the percentage of the total income that was budgeted for each item listed.

1.	Food and Supplies	$188,500	_____ %
2.	Labor	$166,750	_____ %
3.	Workman's Compensation	$ 21,750	_____ %
4.	Fringe Benefits	$ 65,250	_____ %
5.	Utilities	$ 72,500	_____ %
6.	Printing Service	$ 29,000	_____ %
7.	New Equipment	$ 79,750	_____ %
8.	Equipment Service	$ 43,500	_____ %
9.	Telephone	$ 14,500	_____ %
10.	Profit	$ 43,500	_____ %

4. For a food service budget the yearly income from register and supplemental sales is $875,000.00. Find the percentage of the total income that was budgeted for each item listed.

1.	Food and Supplies	$236,250	_____ %
2.	Labor	$183,750	_____ %
3.	Workman's Compensation	$ 43,750	_____ %
4.	Fringe Benefits	$ 70,000	_____ %
5.	Utilities	$ 78,750	_____ %
6.	Printing Service	$ 17,500	_____ %
7.	New Equipment	$105,000	_____ %
8.	Equipment Service	$ 52,500	_____ %
9.	Telephone	$ 26,250	_____ %
10.	Profit	$ 61,250	_____ %

5. The Johnson family has a monthly income of $2,400.00. Their monthly budget is as follows:

Food	$600.00
Clothing	$432.00
Charities	$144.00

Savings	$192.00
Mortgage Payment	$648.00
Recreation	$216.00
Utilities	$168.00

What percent of the monthly income was budgeted for each item? Prove that your percents are correct by adding the results. The sum should be 100%.

6. The Jackson family has a monthly income of $2,650. Their monthly budget is as follows:

Rent	21%
Food	24%
Clothing	16%
Utilities	9%
Entertainment	7%
Insurance	2%
Doctors' Fees	8%
Savings	13%

Find the dollar amount that was budgeted for each item listed. The sum of the completed budget should equal $2,650.00.

7. The Greenwood family has a monthy income of $3,200. Their monthly budget is as follows:

Mortgage Payment	$768.00
Food	$832.00
Clothing	$544.00
Utilities	$288.00
Insurance	$ 96.00
Entertainment	$256.00
Doctors' Fees	$224.00
Savings	$192.00

What percent of the monthly income was budgeted for each item? Prove that your percents are correct by adding the results. The sum should be 100%.

8. The Fahy family has a yearly income of $28,975.00. Their budget for the year is as follows:

| Food | 23% |

Mortgage Payment	27%
Utilities	8%
Clothing	13%
Recreation	7%
Family Welfare	6%
Benevolences	3%
Savings	9%
Other Items	4%

Find the dollar amount that was budgeted for each item listed. The sum of the completed budget should equal the yearly income of $28,975.00.

9. The Curran family has a monthly income of $2,260.00. Their monthly budget is as follows:

Rent	$565.00
Food	$519.80
Clothing	$429.40
Utilities	$180.80
Entertainment	$248.60
Insurance	$ 67.80
Doctors' Fees	$ 90.40
Savings	$158.20

What percent of the monthly income was budgeted for each item? Prove that your percents are correct by adding the results. The sum should be 100%.

10. The Thompson family has a monthly income of $3,450.00. Their monthly budget is as follows:

Rent	30%
Food	21%
Utilities	8%
Clothing	12%
Insurance	2%
Entertainment	9%
Savings	15%
Charities	3%

Find the dollar amount that was budgeted for each item listed. The sum of the completed budget should equal $3,450.00.

PART FIVE

For One and All

In this part of the text, three subjects are discussed that have or will have an effect on the life of every student. They touch one's life in the food service business as well as everyday living. These three subjects are taxes, the metric system, and checking accounts.

Taxes are a form of compulsory contribution, usually monetary, levied by government upon individuals, businesses, and property. The purpose of taxes is to acquire revenue to use for governmental expenses and other public projects. Knowledge of this subject can, in some instances, save money for the individual or business. This subject is therefore covered in Unit 23.

The use of the metric system of measure in the United States is becoming more widespread. Many products sold in grocery stores have both customary and metric units on their labels. Some states have marked their road signs in both miles and kilometers. Many baseball fields have the distances to the fences marked both ways. Several automobile companies have metric programs and many school boards have adopted a "go metric resolution." In order to become proficient in the use of the metric system one must use it as often as possible. As a start, study the information presented in Unit 24 and work the problems.

Business operations and many individuals use a checking account because it is efficient and convenient. Therefore, this subject is covered in Unit 25, and the student should become familiar with this procedure.

UNIT 23
Taxes

Today, people are faced with taxes from three different sources: federal, state, and local (city and county). The following is a list of some of the taxes that restaurant operators and their employees are required to pay, or the employer is required to collect.

Federal:

- Employees Federal Withholding Tax
- Income Tax
- Social Security

State:

- Employees State Withholding Tax
- Income Tax
- Sales Tax
- Various licenses

Local:

- Employees City Withholding Tax
- City Income Tax
- Personal Property Tax
- Occupational Tax
- Various Licenses
- Real Estate Tax

These taxes affect not only the business owner, but each employee, too. Since these taxes affect the employee, and show up in deductions on paychecks, it is important to learn as much as possible about them. In this unit, the goal is to acquaint the student with these taxes and provide enough information to help the student understand their importance.

FEDERAL

Employees Federal Withholding Tax

This is money withheld from each paycheck by an employer during the year, to pay for the income tax that is owed the federal government at the end of the year. The money withheld is sent to the government and credited to the employee's account. Paying the tax as the money is earned is called ''pay as you go.'' At the end of the calendar year (December 31st), and before January 31st of the following year, the employer sends each employee a *W-2 form* (wage and tax statement) showing the total amount of money earned, how much money was withheld for taxes, social security payments, plus any other money that was withheld such as state and city income taxes. If a person has worked steadily for just one employer the whole year, the amount of taxes withheld will probably equal the amount owed at the end of the year. In this case, when the employee files the return (on or before April 15th of the following year), what is owed the federal government is already paid.

There are situations where the employer deducts too little or too much money during the year. If too little money was deducted, and money is owed, a check for the amount owed must be sent with the return. If the amount owed is less than the amount deducted from the paychecks, the government sends the individual a check for the amount overpaid after the return is processed.

The W-2 form received from the employer is in triplicate (3 copies). One copy is to be filed with the federal return, one with the state return, and one should be retained for the individual's personal files. If a single person under age 65 earns less than $4 440 and wishes to get back the money that was withheld by the employer, an income tax return must be filed even though it is not required. This amount is higher for most other filing status. The idea of "pay as you go" withholding tax is to lessen the burden of large payments when taxes become due.

Income Tax

The main source of income for operating the federal government is obtained from taxes levied on the income of individuals and corporations. Income tax returns must be filed by all people who earn over a certain amount of money. (Contact the Internal Revenue Service to find out the exact amount.) These individuals are required by law to file a federal income tax return with the nearest office of the Internal Revenue Service, on or before April 15th.

It is impossible to discuss all of the problems that could arise when preparing a federal income tax return, therefore only the major points are covered. If more information is desired, it can be obtained from a local Internal Revenue Service office. The office will provide assistance in making out returns as well as answer questions. Help can also be obtained from accountants who are licensed by the state to fill out returns for a fee. Remember that although another person may make out your return and sign it, you are also required to sign, and are held responsible for any mistakes.

One of two forms may be used when filling out a return: form 1040A or 1040. The one used depends upon the sources of the income. The short form 1040A can be used if all income is from wages, salaries, tips, other employee compensation, dividends, or interest, and deductions are not itemized. If you do not meet the qualifications for using the short form, the long form 1040 must be used. In most cases, the form needed and the instruction booklet are sent to a person's home. The form sent is based on the form the employee filed the year before.

One concern of food service employees is that income tax must be paid on cash tips. The employee must give the employer a written statement of cash tips if a certain amount in tips is received during a calendar month. The employer deducts from the employee's paycheck the proper amount of withholding tax for these tips, or the employee pays the employer the amount of tax due out of the tips collected.

The cost of meals furnished the employee by the employer is also considered part of the employee's income. It is therefore subject to income tax unless it can be shown that the employee was given the meals for the convenience of the employer.

Social Security Tax

The Social Security Act provides for the payment of retirement, survivor, and disability insurance benefits. It also provides hospital insurance benefits for persons age 65 and over who are eligible (this is commonly known as medicare benefits). Funds for the payment of these benefits are provided by taxes levied upon employees and their employers, and upon self-employed persons.

Under the terms of the Social Security Act, the employer is required by law to deduct a certain percentage of the employee's wages each payday and remit the amounts deducted to the federal government.

The term *wages* means "that which is paid or received for services" and includes salaries, commissions, fees, bonuses, tips, etc.

Example:

Jim Roberts, manager of the Red Gate Restaurant, is paid a salary of $250 per week, plus a 2 percent commission on the party business in excess of $4,000 per week. During the first week of May, the party business was $8,500. Assume that the social security tax rate is 5.85%. How much tax was deducted from Jim's earnings for that week?

Salary	$250.00
Commission	90.00 (2% of $4,500)
Total Earnings	$340.00
5.85% of $340	= $ 19.89

The amount of social security tax deducted from Jim's salary that week was $19.89.

Besides the money deducted each payday from the employee's paycheck, a tax equal to that amount is levied on the employer. The tax is computed at the same rate as the employees tax and is based on the total taxable wages paid by the employer.

Example:

During the month of April the Red Gate Restaurant paid out $5,290 in taxable wages. If the social security rate is 5.85 percent, the owner must pay the federal government $309.47.

$$\$5,290 \times .0585 = \$309.465 \text{ or } \$309.47$$

Self employed persons must also pay social security taxes on income.

STATE

Employees State Withholding Tax

This is money withheld from paychecks by the employer during the year for the purpose of paying the employees state income tax at the end of the year. It is done for the same purpose as the employees federal withholding tax. The amount withheld varies from state to state, but is always lower than the amount withheld for federal tax.

Income Tax

Most states have an income tax. (One exception is the state of Nevada, which gains most of its income from other sources.) The income tax laws in most states are patterned after the

For One and All

federal income tax laws. The amount of tax paid on the adjusted gross income varies from state to state, but it is always less than the amount paid to the federal government.

State Sales Tax

Most states also have a sales tax. This tax is a certain percent of the purchase amount, and is added to the amount of purchase made within that state.

Example:

Bob and Dolores received a dinner check at the Red Gate Restaurant for $18.75. The sales tax for that particular state is 5 percent. How much tax do they pay? What is the total bill?

$18.75   Cost of 2 dinners
× .05   5% Tax
$.9375   = $.94 Amount of tax

$18.75   Cost of 2 dinners
+ .94   Amount of tax
$19.69   Total Bill

To help cashiers and salespeople find the total amount of tax due, a tax chart can be used, showing the amount of purchase and the tax due on that purchase. An example of a state sales tax chart is given in Unit 10.

The business owner is asked to collect this tax and give the receipts to the state government. In some states, a small percentage of the amount collected is given to the business owner as compensation for collecting the tax.

Various Licenses

These vary from state to state, but usually involve licenses to serve food and drink.

LOCAL
Employees City Withholding Tax

This is money withheld from the paycheck by an employer during the year to pay for the income tax that is assessed by the city where the employee works.

City Income Tax

This is money paid to the city to help pay for the cost of operating the city. The money is usually paid to the city in which the employee works, however, this varies from city to city.

Personal Property Tax

This tax is usually assessed by the county in which the employee lives. It is a tax on certain pieces of personal property, such as an automobile. The amount of tax is usually based on the present value of the property. For example, a new car is taxed more heavily than an older one.

Occupational Tax

This tax is assessed by the city in which a business is established. It is a tax that must be paid in order to operate any business within that city.

Various Licenses

These vary from city to city, but in most cases include licenses for food and drink.

Real Estate Tax

This tax is usually assessed by the local governments, which include cities, counties, and townships, to provide the money to pay their schoolteachers, firefighters, police, local officials, etc. The money is also used for local projects such as improving streets and sewers, or building playgrounds. Real estate tax provides most of the revenue needed to operate local governments. It is a tax that affects the property owners most, because they are the ones who benefit most from local services and improvements.

Property taxes are collected once a year or every six months, depending on the area in which the property is located. The tax rate is based on the amount of money required to meet the budget to operate the city, county, school district, and so on, for the next year. Each unit of the local government plans a budget for the following year and sets a tax rate that will meet these expenses. The total of these individual rates makes up a combined rate that is charged to the property owner.

Each piece of property in the area is assessed for its value by the county assessor. The property is usually assessed at 35 to 40 percent of the actual amount it would bring if sold. Some areas assess property at its market value, but this is not usually the case.

The property tax rate may be expressed as a percent, or a certain amount of money for every $100 or $1,000 of assessed valuation.

Example No. 1:

$35,000   Value of house if sold (market value)

$14,000   Assessed valuation (40% of market value)

$32   Tax rate (per $1,000 of assessed value)

$14,000 ÷ $1,000 = $14

$32 × $14 = $448 Cost of real estate taxes for the year

Example No. 2:

$35,000   Value of house if sold (market value)

$14,000   Assessed valuation (40% of market value)

$6.215   Tax rate (per $100 of assessed value)

$14,000 ÷ $100 = $140

$6.215 × $140 = $870.10 Cost of real estate taxes for the year

Note: The tax rate ($6.215 per $100 of assessed value) may also be expressed as 6.215 percent of the total assessed value. The percent is changed to a decimal by moving the decimal point two places to the left when removing the percent sign. This is the same as dividing by 100.

Example No. 3:

$35,000   Value of house if sold (market value)

$14,000   Assessed valuation (40% of market value)

5.5%   Tax rate of the assessed valuation

5.5% × $14,000 = $770

or

.055 × $14,000 = $770 Cost of real estate taxes for the year

ACHIEVEMENT REVIEW

TAXATION

A. Answer questions 1 through 8 on social security tax. Round answers to the nearest cent.

1. Joe Jones, manager of the Sea Shore Restaurant, is paid a salary of $185 per week, plus 1½ percent commission on the party business in excess of $5,000. During the first week of December, the party business amounted to $7,500. If the social security tax rate is 5.85 percent, how much social security tax was deducted from Joe's earnings for that week?

2. The cook at the Sea Shore Restaurant is paid a salary of $168 per week, plus 2 percent commission on all food sales for the week in excess of $10,000. During the second week of March, the food sales amounted to $15,600. If the social security tax rate is 5.85 percent, how much social security tax was deducted from his earnings for that week?

3. The bartender at the Sea Shore Restaurant is paid a salary of $162 per week plus a 1½ percent commission on all bar sales for the week in excess of $8,500. During the last week of April, the bar sales amounted to $10,200. If the social security tax rate is 5.85 percent, how much social security tax was deducted from his earnings for that week?

4. Bill Walter, a waiter at the Red Gate Restaurant, is paid a salary of $80 per week, plus tips. During the first week of April his tips amounted to $90. Assuming that the social security tax rate is 5.85 percent, how much tax was deducted from Bill's earnings for that week?

5. Jean Curran, a waitress at the Red Gate Restaurant, is paid a salary of $80 per week, plus tips. During the first week of May her tips amounted to $103. Assuming that the social security tax rate is 5.85 percent, how much tax was deducted from Jean's earnings for that week?

6. Fred Hartzel, manager of the Kentucky Inn, is paid a salary of $225 per week, plus 2½ percent commission on the food and beverage business in excess of $10,500 per week. During the second week of November, the business amounted to $13,800. Assuming that the

social security tax rate is 5.85 percent, how much tax was deducted from Fred's earnings for that week?

7. During the month of June, the Sea Shore Restaurant paid out $25,650 in taxable wages. If the social security rate is 5.85 percent, find the amount of social security tax the owner must pay.

8. During the month of February, the Red Gate Restaurant paid out $36,225 in taxable wages. If the social security rate is 5.85 percent, find the amount of social security tax the owner must pay.

B. For problems 1 through 10, compute the state income tax due on net income. Use statements a through e as a guide to answer the problems.

a. Net income is $3,000 or less, your tax is 2 percent.
b. Net income is over $3,000, but not over $4,000, your tax is $60 plus 3 percent of the excess over $3,000.
c. Net income is over $4,000, but not over $5,000, your tax is $90 plus 4 percent of the excess over $4,000.
d. Net income is over $5,000, but not over $8,000, your tax is $130, plus 5 percent of the excess over $5,000.
e. Net income is over $8,000 your tax is $280, plus 6 percent of the excess over $8,000.

Problems:
1. Net income is $2,996
2. Net income is $3,995
3. Net income is $12,850
4. Net income is $4,952
5. Net income is $6,780
6. Net income is $16,230
7. Net income is $7,995
8. Net income is $20,600
9. Net income is $4,976
10. Net income is $10,380

C. Find the amount of real estate tax due in problems 1 through 20.

1. The value of the house if sold = $35,000. Assessed valuation is 35 percent of the market value. Tax rate is $34.00 per $1,000 of assessed value. Find the annual taxes due.

2. The value of the house if sold = $42,000. Assessed valuation is 25 percent of the market value. Tax rate is $32.00 per $1,000 of assessed value. Find the annual taxes due.

3. The value of the house if sold = $37,500. Assessed valuation is 32 percent of the market value. Tax rate is $42.00 per $1,000 of assessed value. Find the annual taxes due.

4. The value of the house if sold = $46,875. Assessed valuation is 32 percent of the market value. Tax rate is $36.00 per $1,000 of assessed value. Find the annual taxes due.

5. The value of the house if sold = $50,000. Assessed valuation is 32 percent of the market value. Tax rate is $38.50 per $1,000 of assessed value. Find the annual taxes due.

6. The value of the house if sold = $42,000. Assessed valuation is 30 percent of the market value. Tax rate is $3.255 per $100 of assessed value. Find the annual taxes due.

7. The value of the house if sold = $37,500. Assessed valuation is 36 percent of the market value. Tax rate is $4.358 per $100 of assessed value. Find the annual taxes due.

8. The value of the house if sold = $22,500. Assessed valuation is 40 percent of the market value. Tax rate is $6.225 per $100 of assessed value. Find the annual taxes due.

9. The value of the house if sold = $40,000. Assessed valuation is 36 percent of the market value. Tax rate is $5.325 per $100 of assessed value. Find the annual taxes due.

10. The value of the house if sold = $42,500. Assessed valuation is 36 percent of the market value. Tax rate is $3.954 per $100 of assessed value. Find the annual taxes due.

11. The value of the house if sold = $37,500. Assessed valuation is 32 percent of the market value. Tax rate is 5.241 percent of assessed value. Find the annual taxes due.

12. The value of the house if sold = $22,500. Assessed valuation is 40 percent of the market value. Tax rate is 3.55 percent of assessed value. Find the annual taxes due.

13. The value of the house if sold = $43,000. Assessed valuation is 38

percent of the market value. Tax rate is 4.62 percent of assessed value. Find the annual taxes due.

14. The value of the house if sold = $42,000. Assessed valuation is 30 percent of the market value. Tax rate is 5.72 percent of assessed value. Find the annual taxes due.

15. The value of the house if sold = $38,000. Assessed valuation is 37 percent of the market value. Tax rate is 4.65 percent of assessed value. Find the annual taxes due.

16. The value of the house if sold = $46,000. Assessed valuation is 36 percent of the market value. Tax rate is 5.5 percent of assessed value. Find the annual taxes due.

17. The value of the house if sold = $55,000. Assessed valuation is 35 percent of the market value. Tax rate is 4.55 percent of assessed value. Find the annual taxes due.

18. The value of the house if sold = $54,000. Assessed valuation is 34 percent of the market value. Tax rate is 3.9 percent of assessed value. Find the annual taxes due.

19. The value of the house if sold = $36,500. Assessed valuation is 36 percent of the market value. Tax rate is 4.85 percent of assessed value. Find the annual taxes due.

20. The value of the house if sold = $39,500. Assessed valuation is 35 percent of the market value. Tax rate is 6.5 percent of assessed value. Find the annual taxes due.

UNIT 24
The Metric System

The metric system was introduced to the world by France during the French Revolution. France's lawmakers, during that period of history, asked their scientists to develop a system of measurement based on science rather than custom. They developed a system of measurement which was based upon a length called the meter. A metal bar was used to represent a standard meter of measurement. The meter is slightly longer than one yard.

The metric system is a decimal system based on the number ten. For example, when the meter is divided by ten, it produces 10 decimeters; a decimeter divided by ten produces 10 centimeters; and a centimeter divided by 10 produces 10 millimeters. To put it another way, one meter equals 1 000 millimeters, or 100 centimeters, or 10 decimeters. (Note: The comma is not used in metric notation; instead, a space is left. Ex: 1 000 millimeters.) This system of measure seems more practical when compared to our customary units of measure—the yard, which is divided into 3 feet (or 36 inches); and the foot, which is divided into 12 inches.

The metric system also provides standard rules for amounts of its units through prefixes. For example, a milligram is one thousandth of a gram (weight), a milliliter is one thousandth of a liter(volume), and a millimeter is one thousandth of a meter (length). When the unit is increased and the prefix kilo is added, a kilogram is 1 000 grams and a kilometer is 1 000 meters. The American and English systems lack this kind of uniformity.

Nine-tenths of the world's nations currently use the metric system. If the United States wishes to compete for the world trade, it must adopt the common system used by the rest of the world.

Great Britain and Canada are the newest nations to join the metric world. Both decided on a ten-year transition period. Great Britain announced the start of its ten-year program in 1965 and has already completed the changeover. Canada made its decision in 1971 and has also completed its changeover period.

The United States, however, has found a decision to go metric to be quite a dilemma. For example, in July of 1971, the Department of Commerce recommended that the United States adopt the system in a report made to Congress.

The report stated that the question was not whether the United States should go metric, but how the switch should take place. It proposed that the switch be made over a ten-year period, as in Great Britain and Canada. Little action was taken on this report until December 23, 1975. At that time, President Ford signed into law the Metric Conversion Act, establishing a national policy in support of the metric system and ending the dilemma that had continued for so long.

With the signing of the new law, the dilemma has been resolved through a national policy of coordinating the increasing use of metrics in the United States on a voluntary basis. It also created a Metric Board, appointed by the president, with the advice and consent of the Senate. The board is made up of people from the various economic sectors that will be influenced most by the metric changeover. Included are people representing labor, science, education, small and large business, consumers, manufacturing, construction, and so forth. The function of this board is to plan and carry out a program that will allow the development of a sensible plan for a voluntary changeover.

As the changes take place, people in all types of jobs will be involved, however, the people who will be affected most are those whose jobs are concerned with weights and measures. This is certainly the case for food service workers. Instead of pints, quarts, and gallons, they will have to adjust to liters. Instead of pounds and ounces, they will use kilograms and grams; and instead of degrees Fahrenheit, temperature will be in degrees Celsius (previously known as centigrade). The purpose of this unit, therefore, is to make these terms and others dealing with the metric system more familiar to the food service student.

UNITS OF MEASURE IN THE METRIC SYSTEM

The best way to learn the metric system is to forget all about the customary measurements and think metric. To think metric is to think in terms of ten and to understand the following basics: the meter represents length; grams or in most cases kilograms represent mass; liters, or cubic meters, represent volume; and degrees Celsius deals with temperature. To compare these new units of measure with familiar ones, a meter is about 39 inches, which is slightly longer than the yard. The gram is such a small unit of mass, approximately 0.035 of an ounce, that to make a comparison, it is necessary to take 1000 grams or 1 kilogram, which is equal to 2.2 pounds. The liter is equal to about 1.0567 quarts, which means it is about 5 percent larger than a quart. There are other units of measure in the metric system, but the ones mentioned are those that will be of most concern to the people involved in food service.

Length

It was stated that to think metric is to think in terms of ten. To show how this is done, take the base unit of length (the *meter*) and multiply or divide it by ten. Each time the meter is multiplied or divided by ten, special names are attached on the front of the word to indicate the value. These names are called *prefixes*. For lengths smaller than a meter, it is divided by ten and the result is called a *decimeter*. Dividing a decimeter by ten gives a *centimeter*. When the centimeter is divided by ten, it is called a *millimeter*.

1 decimeter = 0.1 meter
1 centimeter = 0.01 meter
1 millimeter = 0.001 meter

So for the units smaller than a meter, the prefixes are *deci* (a tenth of a meter), *centi* (a hundredth of a meter), and *milli* (a thousandth of a meter).

For lengths larger than a meter, multiply by ten. For ten meters the prefix *deka* is used. For one hundred meters, the prefix is *hecto*. The prefix for one thousand is *kilo*. So ten meters are called a *dekameter*, one hundred meters a *hectometer*, and one thousand meters a *kilometer*. Kilo is a very popular prefix because most distances on roadways are given in kilometers.

1 kilometer = 1000 meters
1 hectometer = 100 meters
1 dekameter = 10 meters

Volume and Capacity

When measuring volume and capacity, it is first necessary to understand what a cubic meter is, before learning what a liter represents. A *cubic meter* is a cube with the sides each one meter long. In other words, a cubic meter equals the length of one meter, the width of one meter, and the height of one meter. If a metal container is one tenth of a meter (one decimeter) on each side, it is referred to as one *liter*. It would contain one liter of liquid and the liquid would weigh one kilogram. When measuring liquid by the American customary system, it is said "A pint is a pound the world around." In the metric system, it can be changed to "A liter is a kilogram the world around," meaning that every liter of liquid weighs one kilogram, or 2.2 pounds. For units smaller than a liter, a container that has sides one centimeter long is called a *cubic centimeter*. It would hold one milliliter of water and one milliliter weighs one gram. From this, of course, it can be seen that in the metric system there is a very direct relationship among length, volume, and mass.

Mass

The base unit for mass is the *gram*, but (as stated before), the gram is such a small unit of weight it did not prove practical for application, so the *kilogram* (1 000 grams) is used as the base unit. It is the only base unit that contains a prefix. When the metric system is adopted, all weights will be given in grams or kilograms. Since the kilogram is a fairly large unit, it may be too large to be a convenient unit for packing most foodstuffs, so the half-kilo (500 grams) may become a more familiar unit. Prefixes such as deci, centi, milli, hecto, and deka may be used with the gram, but they are not practical in everyday life, so the gram and kilogram are the common terms used.

When using the metric system, it has proven difficult to remember the names of all the units and terms, so abbreviations are used, figure 24-1.

In the metric system temperature is measured in degrees Celsius (°C) (previously known as centigrade). On the Celsius scale, the boiling point of water is 100° and the freezing point is at 0°. On the Fahrenheit (F) scale the boiling

Quantity	Unit	Symbol
Length	meter	m
	decimeter	dm
	centimeter	cm
	millimeter	mm
	kilometer	km
	hectometer	hm
	dekameter	dam
Volume	cubic centimeter	cm³
	cubic meter	m³
Capacity	milliliter	ml
	liter	l
Mass	gram	g
	kilogram	kg
Temperature	degrees Celsius	°C

Fig. 24-1 Metric units and their symbols.

point is 212°F and the freezing point is 32°F. See figure 24-2. Actually, the official metric temperature scale is the Kelvin scale, which has its zero point at absolute zero. Absolute zero is the coldest possible temperature in the universe. The Kelvin scale is used often by scientists and very seldom (if ever) used in everyday life.

To convert Fahrenheit temperature to degrees Celsius. Subtract 32 from the given Fahrenheit temperature and multiply the result by ⁵⁄₉.

Example:

$$212°F - 32° = 180°$$

$$\frac{5}{9} \times \frac{180°}{1} = 100°C$$

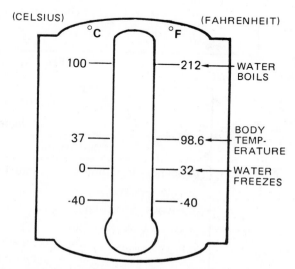

Fig. 24-2 Some common temperatures expressed in Celsius and Fahrenheit.

To convert Celsius degrees to Fahrenheit. Multiply the Celsius temperature by ⅘ and add 32 to the result.

Example:

$$\frac{9}{5} \times \frac{100°}{1} = 180°$$

$$180° + 32° = 212°F$$

Figure 24-3 gives a table of approximate metric conversions. Figure 24-4 is an example of how to use this table to convert a recipe from the customary system to the metric system.

		When you Know	Multiply By	To Find	Symbol
Length		inches	2.5	centimeters	cm
		feet	30	centimeters	cm
		yards	0.9	meters	m
		miles	1.6	kilometers	km
Capacity		teaspoons	5	milliliters	ml
		tablespoon	15	milliliters	ml
		fluid ounces	30	milliliters	ml
		cups	0.24	liters	l
		pints	0.47	liters	l
		quarts	0.95	liters	l
		gallons	3.8	liters	l
Volume		cubic feet	0.03	cubic meters	m^3
		cubic yards	0.76	cubic meters	m^3
Mass		ounces	28	grams	g
		pounds	0.45	kilograms	kg
Temperature		degrees Fahrenheit	$\frac{5}{9}$ (After Subtracting 32)	degrees Celsius	°C

Fig. 24-3 Metric conversions

Fricassee of Veal	
Ingredients	
Customary	**Metric**
18 pounds—Veal shoulder, cut into 1 inch cubes	8.10 kilograms
3 gallons—Water	11.4 liters
2 pounds—Shortening	0.90 kilograms
1 pound 8 ounces—Flour	0.675 grams
Yellow color as desired	
Salt and pepper to taste	

Fig. 24-4 Converting recipe from customary measure to metric measure.

ACHIEVEMENT REVIEW

CONVERTING CUSTOMARY MEASURE TO METRIC MEASURE

A. Convert recipes 1 through 10 from the American customary system to the metric system. Use the conversion table shown in figure 24-3.

1.
White Cream Icing

Ingredients

1 lb. 4 oz. emulsified shortening
1/4 oz. salt
5 oz. dry milk
14 oz. water
5 lb. powdered sugar
vanilla to taste

2.
Yellow Cake

Ingredients

2 lb. 8 oz. cake flour
1 lb. 6 oz. emulsified shortening
3 lb. 2 oz. granulated sugar
1 oz. salt
1 3/4 oz. baking powder
4 oz. dry milk
1 lb. 4 oz. water
1 lb. 10 oz. whole eggs
12 oz. water
vanilla to taste

3.
Italian Meringue

Ingredients

1 lb. egg whites
1 lb. 8 oz. water
1 lb. 12 oz. sugar
1 1/2 oz. egg white stabilizer
1/8 oz. vanilla

4.
Vanilla Pie Filling

Ingredients

12 lb. liquid milk
4 lb. granulated sugar
1 lb. cornstarch
1/4 oz. salt
2 lb. whole eggs
6 oz. butter
vanilla to taste

5.
Fruit Glaze

Ingredients

2 lb. water
2 lb. 8 oz. granulated sugar
8 oz. water
4 oz. modified starch
4 oz. corn syrup
1 oz. lemon juice
food color as desired

6.
Ham Loaf

Ingredients

5 lb. picnic, lean, fresh
8 lb. ham, cured, smoked
1 lb. 8 oz. bread crumbs
1 qt. milk
14 whole eggs, beaten
1 tsp pepper

<table>
<tr><td>

7.

Potato Pancakes

Ingredients

8 lb. red potatoes, peeled
10 oz. onions
8 whole eggs
8 oz. flour
1 oz. salt
1/4 cup parsley chopped
pepper to taste

</td><td>

8.

Bordelaise Sauce

Ingredients

1 lb. onions, minced
1 clove garlic, minced
1 lb. chopped mushrooms
1 gal. brown sauce
1 cup red wine
8 oz. margarine
salt and pepper to taste

</td></tr>
<tr><td>

9.

Swiss Steak

Ingredients

50 round steaks, 6 oz. each
1 lb. onions, minced
2 cloves garlic, minced
12 oz. tomato puree
6 qt. water or brown sauce
3 cups salad oil
12 oz. bread flour
salt and pepper to taste

</td><td>

10.

Soft Dinner Roll Dough

Ingredients

1 lb. granulated sugar
1 lb. 4 oz. hydrogenated
    shortening
8 oz. dry milk
2 oz. salt
6 oz. whole eggs
6 oz. compressed yeast
4 lb. cold water
7 lb. bread flour

</td></tr>
</table>

B. Convert the following Fahrenheit temperatures to Celsius temperatures. Use the formula given in this unit.

1. 180°F
2. 140°F
3. 98°F
4. 210°F
5. 45°F

6. 75°F
7. 120°F
8. 225°F
9. 350°F
10. 400°F

C. Convert the following Celsius temperatures to Fahrenheit temperatures. Use the formula given in this unit.

1. 32°C
2. 204°C
3. 60°C
4. 120°C
5. 20°C

6. 10°C
7. 160°C
8. 140°C
9. 25°C
10. 85°C

UNIT 25
The Checking Account

Today, bills are usually paid by check. This is true for business as well as personal and household expenses. A *check* is a written order, directing the bank to make a payment for the depositor. The bank honors the check and makes the payment, providing the depositor has enough money on deposit in a checking account.

There are four important steps involved in using a checking account.

1. Filling out the deposit slip.

2. Balancing the check register.

3. Writing a check.

4. Checking the bank statement.

If these steps are not done properly, problems result for both the depositor and the bank. These problems can sometimes result in a fine for the depositor. Therefore it is helpful to review the procedure for each step.

THE DEPOSIT SLIP

The deposit slip, figure 25-1, provides the depositor and the bank with a record of the trans-

action when money is deposited in the checking account.

Note that cash and checks are listed separately. After this is done, they are added and the total deposit is shown on the last line. It is extremely important that this total amount be correct.

Once the deposit slip is filled out, the depositor gives the slip, with the deposit, to the bank teller. The teller keeps this slip for the bank's records and gives the depositor another slip called the checking account deposit receipt. On this receipt is stamped the amount deposited in the account and the date.

THE CHECK REGISTER

The *check register*, figure 25-2, is given to the depositor by the bank, so the depositor can record deposits and checks, and thus always know the balance of money on hand. In this way, the depositor always knows the most amount for which a check can be written, and is unlikely to overdraw on the account. (*Overdraw* means to write checks for more than the balance in the account.)

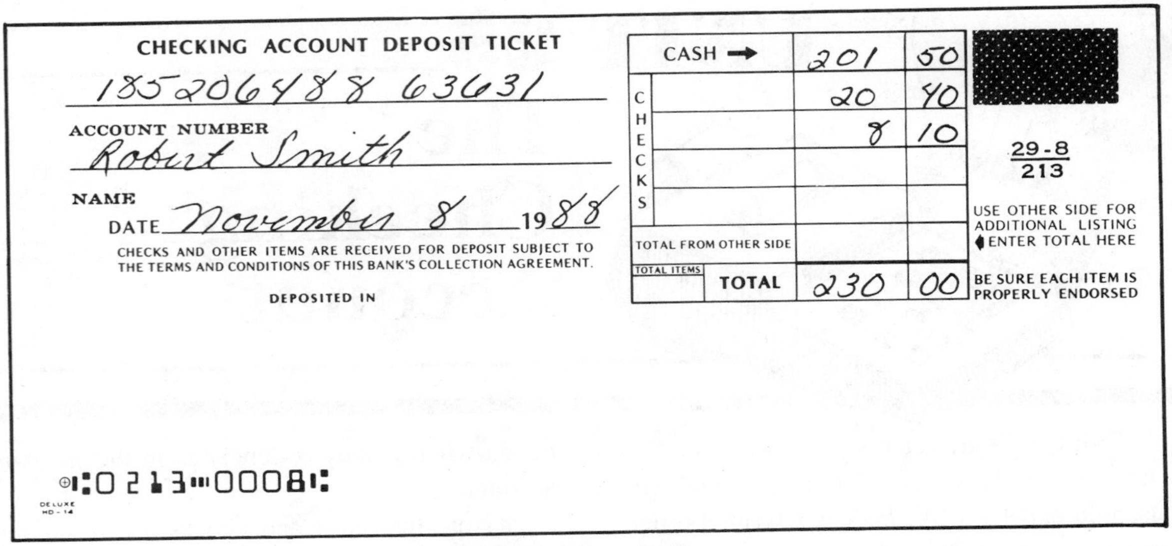

Fig. 25-1 Deposit slip.

The balance brought forward, shown at the top in figure 25-2, is the balance from the previous page in the register. It shows a total of $283. On August 10 a check was drawn for $18, leaving a balance of $265. On August 13, another check was drawn for $175, leaving a balance of $90. On August 16, a deposit of $250 was made. This amount was added to the previous balance, creating a new balance of $340. On August 18 a check for $75 was drawn leaving a balance of $265, and on August 20 another check for $135 was drawn. The remaining balance, $130, is the net amount against which future checks may be drawn.

WRITING A CHECK

The bank issues checks to the depositor which usually come in booklets of 20. The depositor may be required to pay cash for the checks,

or the bank may deduct the cost from the balance. Sometimes the checks are free, if the depositor has a savings account at the same bank.

When writing a check, figure 25-3, page 254, always write neatly and clearly, using ink. Be sure that all of the information listed, such as amount, check number, and date, are correct. Do not forget to sign the check.

THE BANK STATEMENT

At the end of a certain period of time (usually every 3 months, although each bank has its own regulations), the bank provides the depositor with a statement, figure 25-4, page 255. The statement shows the checks drawn and deposits made during that period of time. The bank also returns all checks (cancelled checks) that were issued during that period. The cancelled checks are the depositor's receipts if proof is needed that payment

CHECK NO.	CHECKS DRAWN IN FAVOR OF		DATE	BAL. BRT. FRD.	✓	∥ 283	00
111	TO Cinti Bell		8/10	AMOUNT OF CHECK OR DEPOSIT		18	00
	FOR Telephone Service			BALANCE		265	00
112	TO Allstate Insurance		8/13	AMOUNT OF CHECK OR DEPOSIT		175	00
	FOR Car Insurance			BALANCE		90	00
	TO Deposit		8/16	AMOUNT OF CHECK OR DEPOSIT		250	00
	FOR			BALANCE		340	00
113	TO Shilliton Dept. Store		8/18	AMOUNT OF CHECK OR DEPOSIT		75	00
	FOR Charge Account			BALANCE		265	00
114	TO Norwood Building & Loan		8/20	AMOUNT OF CHECK OR DEPOSIT		135	00
	FOR House Payment			BALANCE		130	00
	TO			AMOUNT OF CHECK OR DEPOSIT			
	FOR			BALANCE			
	TO			AMOUNT OF CHECK OR DEPOSIT			
	FOR			BALANCE			
	TO			AMOUNT OF CHECK OR DEPOSIT			
	FOR			BALANCE			
	TO			AMOUNT OF CHECK OR DEPOSIT			
	FOR			BALANCE			
	TO			AMOUNT OF CHECK OR DEPOSIT			
	FOR			BALANCE			
	TO			AMOUNT OF CHECK OR DEPOSIT			
	FOR			BALANCE			

Fig. 25-2 Check register.

was made. The bank statement is used to check the bank's figures against those recorded in the depositor's check register. In this way, mistakes can be detected before a problem arises or a fine is imposed.

In the top left corner of figure 25-4 is shown the depositor's account number. In the center, the depositor's name and address are given. The balance brought forward shows that as of May 20, there was no money in this account. This

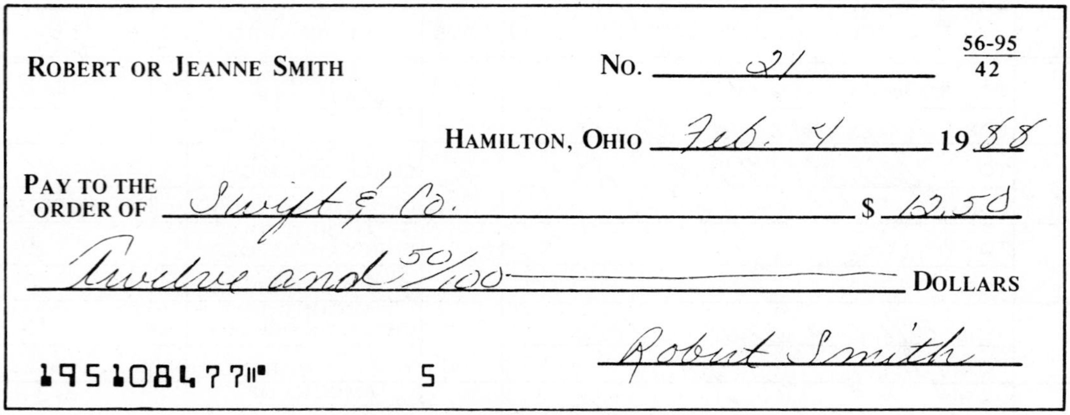
ROBERT OR JEANNE SMITH No. 21 56-95/42

HAMILTON, OHIO Feb. 4 1988

PAY TO THE ORDER OF Swift & Co. $ 12.50

Twelve and 50/100 DOLLARS

1951084770 5 Robert Smith

Fig. 25-3 Writing a check.

was probably when the account was opened at the bank. Deposits and credits show that during this period (May 20 to July 15) 8 deposits were made totaling $1,710.32. This figure can be checked by finding the sum of all the amounts listed with a DP before them. Checks and debits show that 25 checks were drawn on the account and the amount of these checks totaled $1,435.25. There is no service charge indicated by the bank during this period, which shows that the depositor received this service free of charge or paid for the checks when they were received. The current balance shows that as of July 15, $275.07 remained in the account. Other figures shown on this statement include the complete checking activity during this period of time, and dates and amounts of all checks written.

486-174-8		John or Jane Doe
Account Number		9464 Stone Hill Dr.
		Westchester, Ohio 45070

BALANCE FORWARD		DEPOSITS & CREDITS		CHECKS & DEBITS			CURRENT BALANCE	
AS OF	05/20	NO.	AMOUNT	NO.	AMOUNT	SER. CHG.	AS OF	07/15
	00	8	$1,710.32	25	$1,435.25			$275.07

DATE	CHECK NO. OR CODE	AMOUNT	DATE	CHECK NO. OR CODE	AMOUNT	DATE	CHECK NO. OR CODE	AMOUNT
05|20	DP	200|00	07|05		50|00			
06|03	DP	200|00	07|05		139|88			
06|05		100|00	07|08		14|47			
06|07		132|00	07|09		10|50			
06|10	DP	500|00	07|10		14|44			
06|13		105|00	07|10		16|56			
06|13		110|35	07|10		18|40			
06|17	DP	200|00	07|11		50|00			
06|18		20|00	07|11		50|00			
06|19		255|92	07|15	DP	200|00			
06|24		10|00						
06|24		30|00						
06|24		50|00						
06|24		97|00						
06|27	DP	200|00						
06|28		25|16						
07|01	DP	200|00						
07|01		4|13						
07|01		7|00						
07|01		13|25						
07|01		100|00						
07|03	DP	10|32						
07|05		11|19						

Fig. 25-4 Bank statement.

For One and All

ACHIEVEMENT REVIEW
CHECKING ACCOUNT

A.   Prepare deposit slips for problems 1 through 5. Follow the example given in figure 25-1. If deposit slips are not available, make one by listing the necessary information on blank paper.

1.   On August 18, Bill Johnstone deposited the following in his checking account.

2 twenty dollar bills	2 half dollars
4 ten dollar bills	8 quarters
6 five dollar bills	6 dimes
5 one dollar bills	7 nickles
2 checks—$12.75 and $22.25	

2.   On September 8, Charles Curran deposited the following in his checking account.

3 twenty dollar bills	3 half dollars
6 ten dollar bills	5 quarters
4 five dollar bills	6 dimes
3 two dollar bills	8 nickles
8 one dollar bills	
3 checks—$15.90, $25.30, and $42.20	

3.   On October 6, Jim Roberts deposited the following in his checking account.

4 twenty dollar bills	6 half dollars
8 ten dollar bills	3 quarters
7 five dollar bills	7 dimes
4 two dollar bills	9 nickles
9 one dollar bills	
1 check—$53.40	

4.   On October 13, Bill Ford deposited the following in his checking account.

2 twenty dollar bills	3 half dollars
3 ten dollar bills	7 quarters
6 five dollar bills	9 dimes
5 two dollar bills	4 nickles
3 one dollar bills	
2 checks—$26.80 and $35.25	

256

5. On November 10, Bruce Taylor, treasurer of the Cuisine Club, deposited in their checking account the following checks and money collected for dues.

    2 twenty dollar bills   5 half dollars

    2 ten dollar bills   5 quarters

    8 five dollar bills   4 dimes

    6 one dollar bills

    3 checks—$10.40, $8.50, and $6.75

B. Prepare check registers for problems 1 through 5. Follow the example given in figure 25-2. If a check register is not available, make one by listing the necessary information on a sheet of paper.

1. Balance brought forward $520.65

    Oct. 2 Check No. 6 $43.50 Gas and Electric Co.

    Oct. 5 Check No. 7 $25.80 Best State Insurance Co.

    Oct. 10 Deposit $165.50

    Oct. 15 Check No. 8 $35.80 Albers Meat Market

    Oct. 20 Check No. 9 $265.00 Bill's Service Station

2. Balance brought forward $395.35

    Nov. 2 Check No. 10 $14.50 Bell Telephone Co.

    Nov. 6 Check No. 11 $43.50 Gas and Electric Co.

    Nov. 12 Check No. 12 $142.60 Joe's Service Station

    Nov. 14 Deposit $264.25

    Nov. 16 Check No. 13 $15.60 Butler Co. Water Works

    Nov. 17 Deposit $45.20

    Nov. 21 Check No. 14 $352.20 Internal Revenue Service

3. Balance brought forward $680.48

    Dec. 3 Check No. 15 $63.45 Swallens Dept. Store

    Dec. 5 Deposit $183.20

    Dec. 8 Check No. 16 $158.00 Evanston Building and Loan Co.

    Dec. 9 Check No. 17 $178.25 Bay State Insurance Co.

    Dec. 12 Deposit $98.75

    Dec. 15 Check No. 18 $62.78 G.M.A.C.

4. Balance brought forward $728.60

> Jan. 4 Check No. 19 $15.34 Webster Insurance Co.
> Jan. 6 Check No. 20 $187.00 Home Savings and
> Loan Co.
> Jan. 9 Deposit $223.50
> Jan. 12 Check No. 21 $208.60 Joe's Service Station
> Jan. 14 Deposit $197.60
> Jan. 17 Check No. 22 $179.70 McMillians Dept. Store
> Jan. 20 Deposit $368.75
> Jan. 25 Check No. 23 $76.45 Metropolitan Hospital

5. Balance brought forward $1,268.35

> Feb. 6 Check No. 24 $136.25 Robert Bell, M.D.
> Feb. 7 Deposit $89.65
> Feb. 9 Check No. 25 $119.20 State Insurance Co.
> Feb. 11 Check No. 26 $285.75 Bay City Savings
> Feb. 13 Deposit $115.50
> Feb. 14 Check No. 27 $268.90 Hall's Dept. Store
> Feb. 18 Deposit $250.30
> Feb. 23 Check No. 28 $244.36 Internal Revenue Service

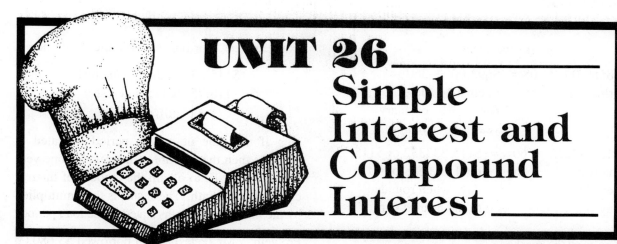

Simple Interest and Compound Interest

Interest means something extra given in return, or money paid for the use of someone else's money for a certain period of time. Interest has become extremely popular in the United States. It can be an asset or a problem, depending on the individual or company doing the borrowing. When done moderately, borrowing can be an advantage; excessive borrowing, however, can cause financial ruin. Many people consider interest a reward for thrift, that is, a payment made to lenders to encourage them to save their money so that money can be made available to others.

There are two types of interest: simple and compound. Simple interest is money paid only on the principal. (The principal is the amount of money loaned or borrowed.) Compound interest is interest that is added to the principal and interest is then paid on the new principal thus obtained. In other words, each time the interest is paid, it is added to the previous principal to obtain a new and higher principal.

The rate of interest paid on a loan is usually expressed as a percentage of the principal (amount borrowed) for a certain period of time, usually a year unless otherwise stated. This percentage interest rate is determined and kept current by various factors. The most important factor is the

relation between the supply of money available to lend and the demand for borrowing. The same factor controls most pricing—the law of supply and demand.

To calculate the amount of interest charged or paid, use the following formula:

Principal $\times$ Rate $\times$ Time = Amount of Interest

Where principal is the amount borrowed, rate is the percentage of principal charged, and the time is usually one year, unless otherwise stated.

The interest year may be expressed as an ordinary year (360 days) or as an exact year (365 days). The type of interest year becomes important when the time of the loan is expressed in days. When this occurs, first find the amount of interest for one year, then divide this amount by either 360 (ordinary) or 365 (exact) days to find the amount of interest per day. The amount of interest charged per day is then multiplied by the number of days mentioned in the loan.

Problem: James Baker borrowed $8,650.00 at 8% interest for a period of 60 days. What amount of interest would he pay if he borrowed the money at (a) ordinary interest? (b) at exact interest?

For part (a) we have

$8650.00 × .08 = $692.00 ÷ 360 =
principal   rate   interest   days
                           per   ordinary
                           year   year

$1.9222222 × 60 = $115.333332 or $115.33
interest   days   total interest for 60
per   days when rounded
day   to the nearest cent

For part (b) we have

$8650.00 × .08 = $692.00 ÷ 365 =
principal   rate   interest   days
                           per   exact
                           year   year

$1.8958904 × 60 = $113.753424 or $113.75
interest   days   total interest for 60
per   days when rounded
day   to the nearest cent

Note: Do not round off the amount of interest per day until it is multiplied by the number of days stated in the loan so that an exact total can be obtained.

The yearly interest for both ordinary and exact interest is the same, but a difference occurs in the amount of interest per day because of the difference in the divisor.

The difference in interest per day will, of course, show a difference in the amount of interest charged for a 60-day period, as shown below. In some cases this difference can be substantial, so be aware of the method used to determine days in a year when borrowing money. Remember, interest calculated on ordinary interest will always be slightly more than that calculated on exact interest. For our example the difference is

$115.33   ordinary interest
113.75   exact interest
1.58   difference

If the time of a loan period is stated in months, then the amount of interest for one year is divided by 12 to obtain the amount of interest charged per month. The result is then multiplied by the number of months stated in the loan period.

Problem: Mr. Pete Jackson borrowed $5,680.00 at 9.5% interest for a period of 6 months. What was the total amount of interest paid?

$5,680.00 × .095 = $539.60 ÷ 12 =
principal   rate   interest   months
                           per year   in year

$44.966666 × 6 = $269.799996 or $269.80
interest   month   total interest for
per month   loan   6 months when
               time   rounded to the
                       nearest cent

Note: Do not round off the amount of interest per month until it is multiplied by the number of months stated in the loan so that an exact total can be obtained.

As stated previously, time is a very important factor in determining the amount of interest paid. It is therefore extremely important that the time factor be calculated accurately. To calculate the exact number of days between dates stated in a loan period, subtract the starting date from the number of days in that particular month and continue to count the days until you have arrived at the ending date.

Problem: How many days are there in a loan

period that starts on May 5th and ends on August 15th?

31	days in month of May
− 5	starting date of loan
26	days left in May
30	days in month of June
31	days in month of July
15	August ending date of loan
102	number of days between starting date and ending date of loan period

Problem: What is the amount of interest charged on a loan of $760.00 at 12.5% ordinary interest if the money is borrowed on April 15th and is to be repaid on September 15th?

Calculate the time

30	days in April
− 15	April starting date
15	days left in April
31	days in May
30	days in June
31	days in July
31	days in August
15	September ending date
153	number of days between starting date and ending date of loan period

Calculate the interest after finding the time

$$\underset{\text{principal}}{\$760.00} \times \underset{\text{rate}}{.125} = \underset{\substack{\text{interest} \\ \text{per year}}}{\$95.00} \div \underset{\substack{\text{days} \\ \text{ordinary} \\ \text{year}}}{360}$$

$$= \underset{\substack{\text{interest} \\ \text{per day}}}{\$.2638888}$$

$$\underset{\text{per day}}{\$.2638888} \times \underset{\substack{\text{number} \\ \text{of days}}}{153} = \underset{\substack{\text{interest for 153 days} \\ \text{when rounded to} \\ \text{the nearest cent}}}{\$40.37}$$

Problem: What is the amount of interest charged on a loan of $2,500.00 at 10.5% exact interest if the money is borrowed on August 12th and is to be repaid on October 26th?

Calculate the time

31	days in August
− 12	August starting date
19	days left in August
30	days in September
26	October ending date
75	number of days between starting date and ending date of loan period

Calculate the interest after finding the time

$$\underset{\text{principal}}{\$2,500.00} \times \underset{\text{rate}}{.105} = \underset{\substack{\text{interest} \\ \text{per year}}}{\$262.50} \div \underset{\substack{\text{days} \\ \text{exact} \\ \text{year}}}{365}$$

$$= \underset{\substack{\text{interest} \\ \text{per day}}}{\$.7119178}$$

$$\underset{\text{per day}}{\$.7119178} \times \underset{\substack{\text{number} \\ \text{of days}}}{75} = \underset{\substack{\text{total interest for} \\ \text{75 days when rounded} \\ \text{to the nearest cent}}}{\$53.393835 \text{ or } \$53.39}$$

Compound interest, as stated previously, is interest payments added to the principal. Interest is then paid on the new principal. The new principal is used each interest period to calculate the amount of interest paid for that period. With interest payments being added on to the principal, the principal will constantly increase, so the amount of interest for each interest period will also increase. This method can be used to let money earn more money.

Interest may be compounded as follows:

- Annually—once a year
- Semiannually—twice a year
- Quarterly—four times a year
- Monthly—twelve times a year
- Daily—every day

When computers became popular, banks and other lending institutions started offering daily compounding on savings accounts. The daily interest earned is usually entered into the account at the end of each quarter as a single amount.

When interest is compounded annually, ordinary simple interest is computed for the first year and then added to the original principal to become the new principal for computing interest for the second year. This procedure is repeated for as many years as stated in the contract.

Problem: Find the amount of compound interest at the end of 3 years on a principal of $2,000.00 compounded annually at 8% interest.

$2000.00 × .08   = $160.00 + $2000.00
original   interest   interest   original
principal   rate   1st year   principal

  = $2160.00
     amount on deposit
     end of 1st year
     new principal

$2160.00 × .08   = $172.00 + $2160.00
new   interest   interest   new
principal   rate   2nd year   principal

= $2332.00
   amount on deposit
   end of 2nd year
   new principal

$2332.00 × .08   = $186.56 + $2332.00
new   interest   interest   new
principal   rate   3rd year   principal

  = $2518.56
     amount on deposit
     end of 3rd year

$2518.56   amount on deposit at the end
           of 3 years
$2000.00   original principal
   518.56 amount of interest earned over
           the 3-year period

Note: Each time the interest is paid, round off to the nearest cent.

When interest is compounded semiannually, the same general procedure explained for compounding annual interest is followed, but the interest is figured twice a year with the time being 6 months or one-half of a year. The total number of periods is twice the number of years stated in the contract period. The interest rate per period would then be half the stated annual rate.

Problem: Find the amount of compound interest at the end of 2 years on a principal of $2,000.00 compounded semiannually at 8% interest.

$2000.00 × .04   = $80.00   + $2000.00
original   interest   interest   original
principal   rate half   6-month   principal
             year   period

= $2080.00
amount on deposit
end of 6 months
(½ year)
new principal

$2080.00 × .04　　= $83.20
new　　　interest　　interest
principal　rate half　2nd 6-month
　　　　　year　　　period

+ $2080.00 = $2163.20
new　　　　amount on deposit
principal　　end of 2nd 6-month
　　　　　　period or 1 year
　　　　　　new principal

$2163.20 × .04　　= $86.528
new　　　interest　　interest
principal　rate half　3rd 6-month
　　　　　year　　　period

+ $2163.20 = $2249.73
new　　　　amount on deposit
principal　　end of 3rd 6-month
　　　　　　period or 1 ½ year
　　　　　　new principal

$2249.73 × .04　　= $89.989
new　　　interest　　interest
principal　rate half　4th 6-month
　　　　　year　　　period

+ $2249.73 = $2339.72
new　　　　amount on deposit
principal　　end of 4th 6-month
　　　　　　period or end of the
　　　　　　2-year period

$2339.72　amount on deposit at the end of
　　　　　2 years with interest compounded
　　　　　semiannually

$2000.00　original principal

$ 339.72　amount of interest earned over a
　　　　　2-year period when compounded
　　　　　semiannually

When interest is compounded quarterly, the interest is figured four times a year or every 3 months. The total number of interest periods is four times the number of years stated in the contract period. The interest rate per period would then be ¼ the stated annual rate.

Problem:　Find the amount of compound interest at the end of one year on a principal of $2,000.00 compounded quarterly at 8% interest.

$2000.00 × .02　　　= $40.00
original　　interest　　interest
principal　　rate　　　1st quarter
　　　　　per quarter

+ $2000.00 = $2040.00
original　　　amount on deposit
principal　　end of 1st quarter
　　　　　　new principal

$2040.00 × .02　　= $40.80　　+ $2040.00
new　　　interest　　interest　　new
principal　rate per　2nd quarter　principal
　　　　　quarter

= $2080.80
amount on deposit
end of 2nd quarter
new principal

$2080.80 × .02　　= $41.616　　+ $2080.80
new　　　interest　　interest　　new
principal　rate per　3rd quarter　principal
　　　　　quarter

= $2122.42
amount on deposit
end of 3rd quarter
new principal

$2122.42 × .02　　= $42.448　　+ $2122.42
new　　　interest　　interest　　new
principal　rate per　4th quarter　principal
　　　　　quarter

= $2164.87
amount on deposit
end of 4th quarter

$2164.87   amount on deposit at the end of one year with interest compounded quarterly

$2000.00   original principal

164.87   amount of interest earned over a 1-year period when compounded quarterly

When interest is compounded monthly, it is figured 12 times a year at $\frac{1}{12}$ of the yearly rate.

When interest is compounded daily, it is figured every day at 1/365 of the yearly rate. In daily compounding, exact simple interest is used rather than ordinary simple interest, which is used when compounding for the other time periods.

Compound interest can be calculated by using the interest formula as shown in the previous examples or by using a compound interest chart (see figure 26-1). The chart is used in many cases to cut down the lengthy process of computing compound interest. It saves time and eliminates confusion. The chart, when used correctly, gives amounts of interest based on one dollar. Figure 26-1 shows a total of 25 periods and a percentage rate from 1% to 7%. The number of periods and the compound interest rates listed in these charts will vary.

Number of Periods	1%	1.5%	2%	2.5%	3%	3.5%	4%	5%	6%	7%
1	1.010000	1.015000	1.020000	1.025000	1.030000	1.035000	1.040000	1.050000	1.060000	1.070000
2	1.020100	1.030225	1.040400	1.050625	1.060900	1.071225	1.081600	1.102500	1.123600	1.144900
3	1.030301	1.045678	1.061208	1.076891	1.092727	1.108718	1.124864	1.157625	1.191016	1.225043
4	1.040604	1.061364	1.082432	1.103813	1.125509	1.147523	1.169859	1.215506	1.262477	1.310796
5	1.051010	1.077284	1.104081	1.131408	1.159274	1.187686	1.216653	1.276282	1.338226	1.402552
6	1.061520	1.093443	1.126162	1.159693	1.194052	1.292255	1.265319	1.340096	1.418519	1.500730
7	1.072135	1.109845	1.148686	1.189686	1.229874	1.272279	1.315932	1.407100	1.503630	1.605781
8	1.082857	1.126493	1.171659	1.218403	1.266770	1.316809	1.368569	1,477455	1.593848	1.718186
9	1.093685	1.143390	1.195093	1.248863	1.304773	1.362897	1.423312	1.551328	1,678479	1.838459
10	1.104622	1.160541	1.218994	1.280085	1.343916	1.410599	1.480244	1.628895	1.790848	1.967151
11	1.115668	1.177949	1.243374	1.312087	1.384234	1.459970	1.539454	1.710339	1.898299	2.104852
12	1.126825	1.195618	1.268242	1.344889	1.425761	1.511069	1.601032	1.795856	2.012196	2.252192
13	1.138093	1.213552	1.293607	1.378511	1.468534	1.563956	1.665074	1.885649	2.132928	2.409845
14	1.149474	1.231758	1.319479	1.412974	1.512590	1.618695	1.731676	1.979932	2.260904	2.578534
15	1.160969	1.250232	1.345868	1.448298	1.557967	1.675349	1.800944	2.078928	2.396558	2.759032
16	1.172579	1.268986	1.372786	1.484506	1.604706	1.733986	1.872981	2.182875	2.540352	2.952164
17	1.184304	1.288020	1.400241	1.521618	1.652848	1.794676	1.947900	2.292018	2.692773	3.158815
18	1.196147	1.307341	1.428246	1.559659	1.702433	1.857489	2.025817	2.406619	2.854339	3.379932
19	1.208109	1.326951	1.456811	1.598650	1.753506	1.922501	2.106849	2.526950	3.025600	3.616528
20	1.220190	1.346855	1.485947	1.638616	1.806111	1.989789	2.191123	2.653298	3.207135	3.869684
21	1.232392	1.367058	1.515566	1.679582	1.860295	2.059432	2.278768	2.785963	3.399564	4.140562
22	1.244716	1.387564	1.545980	1.721571	1.916103	2.131512	2.369919	2925261	3.603537	4.430402
23	1.257163	1.408377	1.576899	1.764611	1.973586	2.206115	2.464716	3.071524	3.819750	4.740530
24	1.269735	1.429503	1.608437	1.808726	2.032794	2.283329	2.563304	3.225100	4,048935	5.072367
25	1.282432	1.450945	1.640606	1.853944	2.093778	2.363245	2.665836	3.386355	4.291871	5.427433

Fig. 26-1 Compound interest chart

Using the compound interest chart is a fairly simple procedure. The steps to take are as follow:

1. Divide the annual interest rate by the number of times the interest will be paid per year. This will give you the percent column to use on the table.

2. Find the total number of interest periods to determine the line to use in the number of periods column.

3. When both columns are determined, place your finger on the number line and move your finger to the right until it comes to the correct rate column. The figure shown where the two columns meet is the compounded interest for $1.00.

4. Multiply this figure or factor by the amount of the original principal to determine the total amount on deposit.

Problem: Jim Kottman deposited $2,000.00 in a savings account paying 6% interest compounded quarterly. What will be (a) the total amount on deposit and (b) the amount of interest earned at the end of 2 years.

Step 1. Find the percent column to use. Divide the annual interest rate by the number of times the interest will be paid in one year.

$$\begin{array}{r} 2\% \text{ percent column to use} \\ \text{Compounded } 4\overline{)\ 8\%} \text{ annual interest rate} \end{array}$$
quarterly

Step 2. Find the total number of interest periods.

2 years $\times$ 4 quarters per year = 8 interest periods

8 is the number to use in the number of periods column.

Step 3. Locate the number 8 in the number of periods column and follow that line to the right to the interest rate column for 2% to find the factor $1.171659.

Step 4. Multiply the factor by the amount of the original principal to determine the amount on deposit at the end of the 2nd-year period.

$1.171659 $\times$ $2000.00 = $2343.318 amount on deposit at the end of 2 years

Thus the answer to part (a) is $2,343.32. For part (b) we write

$$\begin{array}{rl} & \text{total amount on deposit at the} \\ \$2343.32 & \text{end of the 2-year period} \\ -\ 2000.00 & \text{original principal} \\ \hline 343.32 & \text{amount of interest earned over} \\ & \text{a 2-year period at 8\% interest} \\ & \text{per year} \end{array}$$

ACHIEVEMENT REVIEW
SIMPLE INTEREST

1. Find the ordinary simple interest on $575.00 at 6% interest for 2 years.

2. Find the exact simple interest on $985.00 at 7% interest for 2 years.

3. Mr. Joe Curran borrowed $7,540.00 at 8.5% ordinary interest for a period of 60 days. What amount of interest was paid?

4. Mr. Bob Macke borrowed $5,455.00 at 7.5% exact interest for a period of 80 days. What amount of interest was paid?

5. Mr. Charles Oliver borrowed $6,775.00 at 8.5% interest for a period of 8 months. What was the total amount of interest paid?

6. Mr. Bill Kistler borrowed $10,550 at 6.5% interest for a period of 6 months. What was the total amount of interest paid?

7. How many days are there in a loan period that starts on May 9th and ends on September 22nd?

8. How many days are there in a loan period that starts on March 21st and ends on October 20th?

9. What is the amount of interest charged on a loan of $2,800.00 at 10.5% exact interest if the money is borrowed on August 5th and is to be repaid on November 23rd?

10. What is the amount of interest charged on a loan of $880.00 at 12.5% ordinary interest if the money is borrowed on April 20th and is to be repaid on September 20th?

Find the answer to each of the following problems by finding the amount of interest paid using ordinary (360 days) interest.

	Principal	Interest Rate	Time	Interest Paid
11.	$ 960.00	7%	75 days	_____
12.	$1,290.00	9%	110 days	_____
13.	$2,470.00	10.5%	150 days	_____
14.	$3,650.00	12.5%	April 15th to July 10th	_____
15.	$4,225.00	14%	July 5th to Nov. 12th	_____

Find the answer to each of the following problems by finding the amount of interest paid using exact (365 days) interest.

	Principal	Interest Rate	Time	Interest Paid
16.	$ 985.00	6%	60 days	_____
17.	$1,370.00	7.5%	90 days	_____
18.	$2,225.00	11.5%	130 days	_____
19.	$3,845.00	12.5%	April 2nd to July 20th	_____

20. $4,735.00   13%   Aug. 7th to
                                        Dec. 5th   _____

ACHIEVEMENT REVIEW
COMPOUND INTEREST

1. Find the amount of compound interest at the end of 3 years on a principal of $4,000.00 compounded annually at 6% interest.

2. Find the amount of compound interest at the end of 2 years on a principal of $3,550.00 compounded annually at 7.5% interest.

3. Find the amount of compound interest at the end of 2 years on a principal of $1,500.00 compounded semiannually at 9% interest.

4. Find the amount of compound interest at the end of 1½ years on a principal of $4,500.00 compounded semiannually at 10.5% interest.

5. Find the amount of compound interest at the end of 1 year on a principal of $6,000.00 compounded quarterly at 8% interest.

6. Find the amount of compound interest at the end of 2 years on a principal of $9,550.00 compounded quarterly at 5.5% interest.

7. Joe Thomas opened a savings account with a deposit of $1,200.00. The savings and loan paid interest at 6.5% compounded quarterly. If the money was left on deposit for 1 year and 3 months, what was the amount on deposit at the end of 1 year and 3 months? What was the amount of interest earned?

8. Bob Smart purchased a $10,500.00 certificate of deposit paying 10.5% interest compounded monthly for 2 years. What is its total value at maturity? What was the total amount of interest earned?

9. Paul Holden invested $2,250.00 in an IRA account paying 9.5% interest compounded semiannually. What total amount of money will he have on deposit at the end of 4 years? How much interest will he have earned?

10. Carol Fay opened a savings account with a deposit of $6,450.00. The bank paid interest at 7% compounded quarterly. The money was left on deposit for 15 months. What was the amount on deposit at the end of 15 months? What was the total amount of interest earned?

For One and All

For each of the following problems find the amount on deposit at the end of the time period and the amount of interest earned. Use the compound interest table.

	Principal	Interest Rate	Interest Paid	Time in Years	Amount on Deposit	Interest Earned
11.	$ 960.00	6%	Quarterly	1	_____	_____
12.	$ 1,420.00	8%	Semiannually	2	_____	_____
13.	$ 3,890.00	6%	Annually	5	_____	_____
14.	$ 5,175.00	12%	Monthly	1½	_____	_____
15.	$ 8,245.00	12%	Semiannually	7	_____	_____
16.	$ 795.00	10%	Quarterly	6	_____	_____
17.	$16,230.00	5%	Annually	4	_____	_____
18.	$12,448.00	18%	Monthly	2	_____	_____
19.	$ 6,335.00	14%	Semiannually	3	_____	_____
20.	$ 2,990.00	12%	Quarterly	1½	_____	_____

Post-Test: Basic Math Skills

The post-test determines the extent of the student's improvement during the course. In most cases, sufficient improvement will be noted.

To earn a competency in each of the twenty-five math skills presented, a student must work three of the four problems presented correctly. If this is achieved, the student will earn a + (plus) for that particular math skill. If this goal is not achieved, a − (minus) will be recorded. A profile sheet on both the pre- and post-tests are kept on file by the teacher for reference by both the student and the teacher. The pluses earned are recorded in either blue or black ink on the profile sheet under the proper math skill. The minuses are recorded in red ink.

1. Add the following numbers.

$$\begin{array}{r} 29 \\ + 58 \end{array} \qquad \begin{array}{r} 18 \\ 48 \\ 15 \end{array} \qquad \begin{array}{r} 8426 \\ 3268 \\ 6349 \end{array}$$

$49 + 138 + 5,579 + 43,684 =$

2. Subtract the following numbers.

$$\begin{array}{r} 69 \\ - 37 \end{array} \qquad \begin{array}{r} 529 \\ - 263 \end{array} \qquad \begin{array}{r} 73,840 \\ - 12,425 \end{array}$$

$42,229 - 17,483 =$

3. Change the mill to the nearest cent.

$.049 \qquad $6.834

$.137 \qquad $726.533

4. Multiply the following numbers.

$$\begin{array}{r} 43 \\ \times \ 8 \end{array} \qquad \begin{array}{r} 79 \\ \times 36 \end{array} \qquad \begin{array}{r} 883 \\ \times 132 \end{array}$$

$3450 \times 46 =$

5. Divide the following numbers.

$14)\overline{168} \qquad 26)\overline{6864}$

$125)\overline{81,250} \qquad 2232)\overline{618,264}$

6. Reduce the fractions to the lowest terms.

$$\frac{3}{12} \qquad \frac{18}{96} \qquad \frac{54}{108} \qquad \frac{80}{128}$$

7. Convert each mixed number to an improper fraction.

$7\frac{1}{8} = \qquad 18\frac{4}{7} =$

$5\frac{3}{8} = \qquad 45\frac{7}{8} =$

8. Convert each improper fraction to a whole number or to a mixed number.

$\frac{45}{9} \qquad \frac{176}{110}$

$\frac{32}{6} \qquad \frac{280}{22}$

9. Find the equivalent fractions.

$\frac{5}{9} = \frac{?}{54} \qquad \frac{5}{8} = \frac{?}{40}$

$\frac{2}{3} = \frac{?}{18} \qquad \frac{3}{16} = \frac{?}{32}$

10. Add the following fractions and reduce the answer to the lowest terms.

$$\dfrac{7}{8} \qquad 3\dfrac{3}{8} \qquad 1\dfrac{5}{16} \qquad 3\dfrac{1}{2}$$

$$+\dfrac{5}{8} \qquad +4\dfrac{3}{4} \qquad +6\dfrac{1}{4} \qquad 5\dfrac{5}{16}$$

$$+6\dfrac{5}{8}$$

11. Subtract the following fractions and reduce the answer to the lowest terms.

$$\dfrac{11}{12} \qquad 7 \qquad 14\dfrac{1}{4} \qquad 36\dfrac{21}{32}$$

$$-\dfrac{5}{12} \qquad -2\dfrac{3}{8} \qquad -6\dfrac{3}{16} \qquad -24\dfrac{3}{16}$$

12. Multiply the following fractions and reduce the answer to the lowest terms.

$$\dfrac{2}{5} \times \dfrac{5}{6} = \qquad 6\dfrac{3}{4} \times 2\dfrac{7}{8} =$$

$$7 \times \dfrac{5}{9} = \qquad 9\dfrac{2}{3} \times 3\dfrac{1}{4} =$$

13. Divide the following fractions and reduce the answer to the lowest terms.

$$\dfrac{5}{9} \div \dfrac{1}{4} = \qquad 4\dfrac{3}{4} \div \dfrac{4}{5} =$$

$$3\dfrac{1}{4} \div \dfrac{1}{8} = \qquad 15\dfrac{3}{4} \div 1\dfrac{5}{8} =$$

14. Convert each decimal to a fraction.

.8   .008

4.25   .86

15. Convert each fraction to a decimal.

$$\dfrac{3}{16} \qquad\qquad 6\dfrac{1}{6}$$

$$\dfrac{5}{8} \qquad\qquad 2\dfrac{3}{8}$$

16. Convert each fraction to a percent.

$$\dfrac{4}{5} \qquad\qquad \dfrac{7}{9}$$

$$2\dfrac{3}{5} \qquad\qquad \dfrac{7}{10}$$

17. Convert each decimal to a percent.

.0076   5.3

.63   .074

18. Convert each percent to a decimal.

9.6%   14.8%

223%   24%

19. Add the decimals.

$$\begin{array}{r} 8.5 \\ 3.7 \\ +\ .9 \\ \hline \end{array} \qquad \begin{array}{r} 7.096 \\ +\ 15 \\ \hline \end{array}$$

.5 + 7.4 + 14 =

.0036 + .035 + 4.26 =

20. Subtract the decimals.

$$\begin{array}{r} 8.09 \\ -\ 2.68 \\ \hline \end{array} \qquad \begin{array}{r} 245.08 \\ -\ 7.4 \\ \hline \end{array}$$

44.6 − 10.53 =

6 − .07 =

21. Multiply the decimals.

     .295   8.35
  $\times$ 18 $\times$ 2.5

$7.5 \times .035 =$

$12.75 \times .065 =$

22. Divide the decimals.

  $.08\overline{)\ .956}$ $7.5\overline{)\ .5842}$

  $.43\overline{)\ 5453}$ $18\overline{)\ 9.762}$

23. Use percents to find the total number.

  25% of what number is 9?
  15% of what number is 36?
  70% of what number is 56?
  35% of what number is 105?

24. Find the percent of a number.

  38% of 20 =
  7.5% of 48 =
  15½% of 95 =
  85% of 220 =

25. Find the percent of two given numbers.

  What percent of 60 is 36?
  What percent of 90 is 54?
  What percent of 120 is 36?
  What percent of 260 is 91?

Basic Math Skills—Profile Sheet

Student's Name	Addition	Subtraction	Mills	Multiplication	Division	Reducing fractions	Converting to improper fraction	Improper fractions to mixed number	Finding equivalent fractions	Addition of fractions	Subtraction of fractions	Multiplication of fractions	Division of fractions	Converting decimals to fractions	Converting fractions to decimals	Converting fractions to percent	Converting decimals to percent	Converting percent to decimals	Addition of decimals	Subtraction of decimals	Multiplication of decimals	Division of decimals	Percents to find number	Finding percent of a number	Finding percent of two given numbers
1.																									
2.																									
3.																									
4.																									
5.																									
6.																									
7.																									
8.																									
9.																									
10.																									
11.																									
12.																									
13.																									
14.																									
15.																									
16.																									
17.																									
18.																									
19.																									
20.																									
21.																									
22.																									
23.																									
24.																									
25.																									

+ Mastered
– Not Mastered

Glossary

Accounts payable—money owed by the business operator for purchases made.

Accounts receivable—money that is owed to the business operator by customers.

Adhere—to stick fast; to become attached or cling to.

A la carte—foods ordered and paid for separately; usually prepared to order.

Annual—pertaining to a year; happening once in twelve months.

Approximate—to come near to; nearly correct.

Aspic—a clear meat, fish, or poultry jelly.

Assess—to fix or determine the amount of a tax, fine, or damage; to rate or set a certain charge upon, as for taxation.

Assessor—a person appointed to estimate the value of property for the purpose of taxation.

Assets—things of value; all the property of a person, company, or estate that may be used to pay debts.

Balance—difference between the debit and credit sides of an account; an amount remaining over.

Balance sheet—a written statement made to show the true financial condition for a person or business by exhibiting assets, liabilities or debts, profit and loss, and net worth.

Bank note—a note issued by a bank that must be paid to the bearer upon demand. Bank notes are used as money.

Breading—a process of passing an item through flour, egg wash (egg and milk), and bread crumbs, before it is fried.

Budget—a plan of systematic spending; to plan one's expenditures of money, time, etc.

Calculate—to reach a conclusion or answer by a reasoning process.

Calculator—one who computes; a machine that does automatic computations.

Calendar year—a period that begins on January 1 and ends on December 31; consisting of 365 days, in a leap year 366 days.

Capacity—power of holding or grasping; room; volume; power of mind; character; ability to hold cubic content.

Capital—amount of money or property that a person or company uses in carrying on a business.

Celsius—a term used to measure temperature in the metric system of measuring; graduated or divided into 100 equal parts called degrees; previously called centigrade.

Centigrade—a term used to measure temperature in the metric system of measuring; graduated or divided into 100 equal parts called degrees. The term now used is Celsius.

Centimeter—the one hundredth part of a meter.

Certificate—issued by a bank to a depositor indicating that a specific amount of money is set aside and not subject to withdrawal except on surrender of the certificate, usually with an interest penalty.

Chaud-froid—jellied white sauce, used for decorating certain foods that are to be displayed.

Cipher—zero.

Commission—pay based on the amount of business done.

Compensation—something given in return for a service or a value.

Competency—the state of being fit or capable.

Compound—composed of more than one part.

Compressed—made smaller by applying pressure.

Concept—a mental idea of a class of objects.

Constant—always the same; not changing.

Convert—change; to turn the other way around.

Corporation—a group of persons who obtain a charter giving them (as a group) certain legal rights and privileges distinct from those of the individual members of the group.

Currency—money in actual use in a country. In the United States the term usually applies to paper money although technically it is both coins and paper money.

Debit—the entry of an item in a business account showing something owed or due.

Decimal—a system of counting by tens and powers of ten. Each digit has a place value ten times that of the next digit to the right.

Decimeter—a metric measure of length equal to one tenth of a meter.

Deduction—the process of taking away.

Dekameter—a metric measure of length equal to 10 meters.

Denominator—number below the line in a fraction, stating the size of the parts in relation to the whole.

Digit—any one of the figures 0-1-2-3-4-5-6-7-8-9.

Dividend—(1) money to be shared by those to whom it belongs. If a company shows a profit at the end of a certain period, it declares a dividend to the owners of the company. (2) Also, the number to be divided by the divisor.

Divisor—a number by which another (the dividend) is divided.

Economic—pertaining to the earning, distributing, and using of wealth and income, public or private.

Entree—the main dish of a meal.

Equation—to make equal.

Equivalent—equal in value or power.

Estimate—a judgment or opinion in determining the size, value, etc. of an item.

Evaluate—find the value or amount of; fix the value.

Expenditure—that which is spent.

Fabricated—made up; in food service, standardized portions.

Factor—one of the two or more quantities which when multiplied together, yield a given product; example: 2 and 4 are factors of 8.

File—put away and kept in an easy-to-find order.

Finances—money; funds; revenues; financial condition.

Financial—having to do with money matters.

Fiscal year—the time between one yearly settlement of financial accounts and another. In the United States, a fiscal year usually starts July 1 and ends June 30 of the following year.

Fixed assets—those assets (things of value) that stay firm and do not change.

Fixed costs—those costs (price paid) that stay firm and will not change.

Fluctuate—change continually.

Forecast—a prophecy of prediction.

Formula—a rule for doing something; a recipe or prescription.

Fraction—one or more of the equal parts of a whole; a small part or amount.

Function—a quantity the value of which varies with that of another quantity.

Garnish—to decorate, such as food.

Gelatin—an odorless, tasteless substance obtained by boiling animal tissues. It dissolves easily in hot water and is used in making jellied desserts and salads.

Gourmet—a lover of fine foods.

Gram—metric system unit of weight (mass). Twenty-eight grams equal one ounce.

Gratuity—a present of money given in return for a service; also called a tip.

Gross—with nothing removed or taken out. Gross receipts are all the money taken in before costs are deducted.

Gross margin—sales less the cost of food gives the gross margin. It is the margin before other deductions are taken.

Hectometer—measure of length in the metric system equal to 100 meters.

Horizontally—parallel to the horizon; at right angles to a vertical line.

Hypothetical—something assumed or supposed.

Indicator—one who or that which points out.

Ingredient—one part of a mixture.

Installment—part of a sum of money or debt to be paid at certain regular times.

Interest—money that is paid for the use of money.

Inventory—a detailed list of items with their estimated value.

Invert—to turn upside down.

Invoice—a list of goods sent to a purchaser showing prices and amounts.

Itemize—to state by items, as to itemize a bill.

Kilogram—a metric measure of weight equal to 1 000 grams.

Kilometer—a measure of length in the metric system equal to 1 000 meters.

Lease—a written contract whereby one party grants to another party the use of land, buildings, or personal property, for a definite consideration known as rent, for a specified term.

Legumes—vegetables; also refers to dried vegetables such as beans, lentils, and split peas.

Liability—a state of being under obligation; responsible for a loss, debt, penalty, or the like.

Liter—a measure of volume in the metric system. One liter equals 1.0567 quarts in customary liquid measure or 0.908 quarts in dry measure.

Markup—marked for sale at a higher price.

Meringue—egg whites and sugar beaten together to form a white frothy mass; used to top pies and cakes.

Meter—unit of length in the metric system equal to 39.37 inches.

Metric—pertaining to the meter or the system of measurements based on it.

Minuend—number or quantity from which another is to be subtracted.

Mortgage—claim on property given to a person who has lent money in case the money is not repaid when due.

Multiple—a number that contains another number a certain amount of times, without a remainder; example: 16 is a multiple of 4.

Multiplicand—in multiplication, the number or quantity to be multiplied by another number called the multiplier.

Multiplier—number by which another number is to be multiplied.

Net worth—excess value of resources over liabilities; also called net assets.

Numeral—symbol for a number.

Numerator—number above the line in a fraction which shows how many parts of the denominator are taken.

Pasta—a dried flour paste product; examples: spaghetti, vermicelli, and lasagna.

Percent—rate or proportion of each hundred; part of each hundred.

Periodical—magazine that is published regularly.

Perpetual—continuous, endless.

Physical inventory—a count taken of all stock on hand.

P and L sheet—referring to a profit and loss statement.

Portion—a part or share.

Prefix—a letter, syllable or group of syllables placed at the beginning of a word to modify or qualify its meaning; example: *deci* in front of meter indicates $\frac{1}{10}$ of a meter.

Procedure—a way of proceeding; method of doing a task.

Profile—an outline or contour.

Proprietorship—ownership.

Quantity—amount; how much.

Quotient—number obtained by dividing one number by another; the final answer of a division problem.

Ratio—the quotient of one quantity divided by another of like kind, usually expressed as a fraction; example: when baking rice the ratio is 2 parts liquid to 1 part rice—this means to use twice as much liquid as rice.

Receipt—a written statement that money, a package, a letter, etc. has been received.

Reconstitute—to rebuild the way it was originally, to put back into original form; example: to reconstitute dried milk, the water is put back.

Report—an account officially expressed, generally in writing.

Requisition—a demand made, usually in written form, for something that is required.

Rotating menu—menu that alternates by turn in a series. The series is usually set up on a yearly basis.

Roux—a thickening agent consisting of equal parts of flour and shortening.

Sales revenue—money coming in from the sale of certain items.

Sauté—to cook in shallow grease.

Specification—a detailed statement of particulars.

Standardize—to make standard in size, shape, weight, quality, quantity, etc.

Status—condition, state, or position.

Stockholder—owner of stocks or shares in a company.

Subtrahend—number or quantity to be subtracted from another.

Sum—total of two or more numbers or things taken together; the whole amount.

Summarize—express briefly; give only the main points.

Symbol—something that stands for or represents something else.

Table d'hote—a meal of several courses served at a set price. The dinner menu in most restaurants is served table d'hote.

Triplicate—to make threefold; three identical copies.

Unit—a single thing.

Utilities—companies that perform a public service. Railroads, gas and electric, and telephone companies are utilities.

Variable costs—costs that are changeable.

Variation—the extent to which a thing changes.

Vertical—straight up and down.

Volume—space occupied.

Voucher—a written evidence of payment; receipt.

Yield—amount produced.

Index